The Nature of Design

THE
NATURE
of DESIGN

Ecology, Culture, and Human Intention

David W. Orr

OXFORD
UNIVERSITY PRESS

OXFORD
UNIVERSITY PRESS

Oxford   New York
Auckland   Bangkok   Buenos Aires   Cape Town   Chennai
Dar es Salaam   Delhi   Hong Kong   Istanbul   Karachi   Kolkata
Kuala Lumpur   Madrid   Melbourne   Mexico City   Mumbai   Nairobi
São Paulo   Shanghai   Taipei   Tokyo   Toronto

Copyright © 2002 by Oxford University Press, Inc.

First published in 2002 by Oxford University Press, Inc.
198 Madison Avenue, New York, New York 10016

www.oup.com

First issued as an Oxford University Press paperback, 2004.

Oxford is a registered trademark of Oxford University Press

Library of Congress Cataloging-in-Publication Data
Orr, David W., 1944–.
The nature of design : ecology, culture, and human intention / by David W. Orr.
    p.   cm.
Includes bibliographical references (p.).
ISBN-13 978-0-19-514855-8; 978-0-19-517368-0
ISBN 0-19-514855-X; 0-19-517368-6 (pbk.)
1. Nature—Effect of human beings on.
2. Human ecology—Moral and ethical aspects.
3. Environmental responsibility.
4. Global environmental change.
I. Title.
GF75 .O77   2002
304.2'8—dc21   2001036413

We gratefully acknowledge permission from Blackwell Science, Ltd., to reprint
in this book, in somewhat altered form, material from the following articles by
David W. Orr that were originally published in *Conservation Biology*: "Technological Fundamentalism" (8:2, June 1994); "Twine in the Baler" (8:4, December 1994);
"Conservatism and Conservation" (9:2, April 1995); "None So Blind" (9:5, October
1995); "Slow Knowledge" (10:3, June 1996); "Architecture as Pedagogy II" (11:3,
June 1997); "Speed" (12:1, February 1998); "The Limits of Nature and the Nature
of Limits" (12:4; August 1998); "The Architecture of Science" (13:2, April 1999);
"Verbicide" (13:4, August 1999); "Education, Careers, Callings" (13:6, December
1999); "2020: A Proposal" (14:2, April 2000); "Ideasclerosis I" (14:4, August 2000);
Ideasclerosis II" (14:6, December 2000).

9 8 7 6 5 4

Printed in the United States of America
on recycled, acid-free paper

For Wil

Acknowledgments

For the past six years, ecological design has been more than an abstraction for me. The essays that follow originated in physical and intellectual proximity to an ecological design project on the campus of Oberlin College described in chapter 14. What began as a fairly straightforward design and construction project became a crash course in architecture, engineering, materials analysis, ecological engineering, landscape ecology, energy analysis, philosophy, institutional politics, and fund-raising. During that time it was my privilege to work with some of the most remarkable designers of our time. To all of the participants in that project I owe a large debt: Ray Anderson, David Austin, Bill Browning, Kevin Burke, Leo Evans, Carol Franklin, Chris Hays, Mark Hoberecht, Amory Lovins, John Lyle, Bill McDonough, Dave Nelson, Ron Perkins, Russell Perry, Mark Rusitsky, Bob Scheren, Michael Shaw, Stephen Strong, John Todd, Martin Troutman, and Adrian Tuluca. They persevered on a tough project. I owe a large debt to friends and colleagues here and elsewhere particularly David Benzing, Peter Buckley, Fritjof Capra, Tony Cortese, Nancy Dye, Karen Florini, Dierdre Holmes, Jon Jensen, Adam Lewis, Peter Lewis, Al MacKay, Brad Masi, Gene Matthews, Carl McDaniel, John

Petersen, John Powers, Michael Stranahan, Paige Wiegman, and Cheryl Wolfe. I thank David Ehrenfeld and Gary Meffe, both for their editorial skill that improved many of these essays and for their encouragement. I am grateful to Island Press for permission to include "The Ecology of Giving and Consuming" from *Consuming Desires*, ed. Roger Rosenblatt (1999); to MIT Press for permission to include "Loving Children" from *Children and Nature: Psychological, Sociocultural, and Evolutionary Investigations*, ed. Peter H. Kahn Jr. and Stephen R. Kellert (2002); to *Wild Earth* for permission to include "The Great Wilderness Debate, Again"; and to Blackwell Science for permission to reprint material from *Conservation Biology* included here in chapters 3–17. Finally, this book is dedicated to my brother Wilson, with gratitude and love.

Contents

I. The Problem of Ecological Design

1. Introduction: The Design of Culture and the
    Culture of Design   3
2. Human Ecology as a Problem of Ecological Design   13

II. Pathologies and Barriers

3. Slow Knowledge   35
4. Speed   43
5. Verbicide   53
6. Technological Fundamentalism   61
7. Ideasclerosis   68
8. Ideasclerosis, Continued   75

III. The Politics of Design

9. None So Blind: The Problem of Ecological Denial (*with David Ehrenfeld*)   85
10. Twine in the Baler   91
11. Conservation and Conservatism   97
12. A Politics Worthy of the Name   104
13. The Limits of Nature and the Educational Nature of Limits   118

IV. Design as Pedagogy

14. Architecture and Education   127
15. The Architecture of Science   135
16. 2020: A Proposal   143
17. Education, Careers, and Callings   152
18. A Higher Order of Heroism   160

V. Charity, Wildness, and Children

19. The Ecology of Giving and Consuming   171
20. The Great Wilderness Debate, Again   187
21. Loving Children: The Political Economy of Design   198

Bibliography   221

Index   233

§ 1

THE PROBLEM OF ECOLOGICAL DESIGN

1

Introduction: The Design of Culture and the Culture of Design

Environmentalists are often regarded as people wanting to stop one thing or another, and there are surely lots of things that ought to be stopped. The essays in this book, however, have to do with beginnings. How, for example, do we advance a long-delayed solar revolution? Or begin one in forest management? Or materials use? How do we reimagine and remake the human presence on earth in ways that work over the long haul? Such questions are the heart of what theologian Thomas Berry (1999) calls "the Great Work" of our age. This endeavor is nothing less than the effort to harmonize the human enterprise with how the world works as a physical system and how it ought to work as a moral system. In the past two centuries the human footprint on earth has multiplied many times over. Our science and technology are powerful beyond anything imagined by the confident founders of the modern world. But our sense of proportion and depth of purpose have not kept pace with our merely technical abilities.

Our institutions and organizations still reflect their origins in another time and in very different conditions. Incoherence, disorder, and violence are the hallmarks of the modern world. If we are to build a better world—one that can be sustained ecologically and one that sustains us spiritually—we must transcend the disorder and fragmentation of the industrial age. We need a perspective that joins the hard-won victories of civilization, such as human rights and democracy, with a larger view of our place in the cosmos—what Berry calls "the universe story." By whatever name, that philosophy must connect us to life, to each other, and to generations to come. It must help us to rise above sectarianism of all kinds and the puffery that puts human interests at a particular time at the center of all value and meaning. When we get it right, that larger, ecologically informed enlightenment will upset comfortable philosophies that underlie the modern world in the same way that the Enlightenment of the eighteenth century upset medieval hierarchies of church and monarchy.

The foundation for ecological enlightenment is the 3.8 billion years of evolution. The story of evolution is a record of design strategies as life in all of its variety evolved in a vast efflorescence of biological creativity. The great conceit of the industrial world is the belief that we are exempt from the laws that govern the rest of the creation. Nature in that view is something to be overcome and subordinated. Designing with nature, on the other hand, disciplines human intentions with the growing knowledge of how the world works as a physical system. The goal is not total mastery but harmony that causes no ugliness, human or ecological, somewhere else or at some later time. And it is not just about making things, but rather remaking the human presence in the world in a way that honors life and protects human dignity. Ecological design is a large concept that joins science and the practical arts with ethics, politics, and economics.

In one way or another all of the important questions of our age have to do with how we get on with the Great Work, transforming human activity on the earth from destruction to participation and human attitudes toward nature from a kind of autism to a competent reverence. It would be foolish to think that what has taken several centuries or longer can be undone quickly or even entirely. But it would also be the height of folly to continue on our present course or to conclude that we are doomed and give up hope. For most of us the Great Work must begin where we are, in the small acts of everyday life, stitching together a pattern of loyalty and faithfulness to a higher order of being. The hallmarks of those engaged in Great Work everywhere must be largeness of heart, breadth of perspective, practical competence, moral stamina, and the kind of intelligence that discerns ecological patterns.

This is a tall order, but we have a heritage of ecological design intelligence available to us if we are willing to draw on it. The starting point for ecological design is not some mythical past, but the heritage of design intelligence evident in many places, times, and cultures prior to our own. We don't need to reinvent wheels. What we will need in the decades ahead is to rediscover and synthesize, as well as invent. Let me illustrate with four examples.

1. Several days after the bombing of the Murrah Federal Building in Oklahoma City in 1995, an Amish friend of mine with a well-developed sense of humor called from a pay phone to inform me that no Amish person was involved in the crime. I responded by saying that I was not particularly surprised. "Good," he replied, "I just wanted to clarify that in your mind." After a pause he added: "You know if the Amish were involved, the getaway buggy would have been blown up."

My friend usually has a point to make. This time it was simply a humorous way of saying that if the horse is your primary mode of transportation, there are some things you cannot do. Whatever malice may be hidden in the heart, the speed and power of the horse sets limits to the havoc one can cause. If the horse is your primary form of transportation, you cannot haul enough diesel and fertilizer to blow up large buildings, and you could not escape the ensuing destruction anyway. A horse-drawn buggy has a radius of about eight miles in hilly country, and if you have chores to finish by suppertime, you cannot conveniently shop until you drop. And if you could, you still could not haul it all home. The use of draft animals also limits the amount of land one can farm, which, in turn, limits the desire to take over a neighbor's farm.

In Amish culture, in other words, the horse functions like a mechanical governor on a machine. The horse sets a standard of sorts for human activity and a way for the culture to say no to some possibilities, which means saying yes to better ones. The Amish voluntarily accept the limits imposed by the horse and the discipline of living in a close-knit community. People in industrial culture, on the other hand, have no functional equivalent of the horse and accept few limits beyond those of what is assumed to be cost-effectiveness. The Amish and most traditional cultures can sustain themselves indefinitely within the ecological limits of their regions. They contribute little or nothing to climatic change, cancer rates, and the loss of biodiversity, and they are invulnerable to any technological failure originating within their own community. Modern societies, on the other hand, are increasingly vulnerable to a long list of ecological, economic, technological, and social threats. The question then arises whether we also need some functional equivalent of the horse in order to become sustainable. If so, what could it be?

2. The hamlet of Harberton, with a population of perhaps 100, is no more than 4 miles from the city of Totnes (Devon, U.K.) with a population of 10,000. The road connecting the two, however, is a single lane flanked by high hedgerows which traverses an ancient and competently used countryside. Drivers meeting on the lane connecting Harberton and Totnes must decide who will back up to let the other through. The process works with a civility and friendliness that is surprising to an American driver accustomed to speed and rudeness. In fact, the entire scene is unexpected. In, say, Ohio, there would be little or no countryside between the two places. Developers would have filled the four miles with malls, scenic motels, billboards, parking lots, fast-food joints, and poorly constructed housing. In contrast, the people of Devon have maintained and in some ways have improved a landscape continuously inhabited since the Neolithic era. It is a landscape of rolling hills, stone buildings clustered into villages, small fields, dairy farms, sheep pastures, hedgerows, and narrow roads. To the north is an expanse known as Dartmoor, to the south is the English channel and port towns such as Dartmouth from which the Mayflower sailed. This was an ancient landscape before the birth of Socrates and would still be mostly familiar to its early inhabitants. How is it that human occupation and use of this land for perhaps 10,000 years has not led to its desecration?

3. Western agriculture imposed on the island of Bali displaced an agricultural system of remarkable productivity that had thrived for a thousand years or more. Balinese agriculture was controlled by a system of temples presided over by a priesthood that orchestrated the distribution of irrigation water. The entire process was calibrated to the seasons, pests, and differing crop needs by a complex

calendar worked out over many centuries. That intricate, resilient, and highly productive system was displaced by the Green Revolution in the 1970s administered by experts who regarded agriculture as merely technical. The results were disastrous. Crops failed, pests multiplied, and the society unraveled. The Balinese system of agriculture had been a remarkable blend of religion with hydrological and biological management. The imposition of technocratic Western agriculture undid in a few years what had taken hundreds of years to create largely because "the managerial role of the water temples was not easily translated into the language of bureaucratic control" (Lansing 1991, 127). Now much of that system based on Western science and agronomy has been dismantled. But how can a system based on superstition work where one purportedly based on science does not?

4. Designer Victor Papenek once identified the Inuit people of northern Alaska as the best designers in the world. They are, he believed, "forced into excellence by climate, environment, and their space concepts. At least equally important is the cultural baggage they carry with them" (Papenek 1995, 223). Living in spare environments frozen through much of the year, the Inuit people have had to develop acute powers of observation, memory, and senses. They can repeat a long trek using nothing more than the memory of the same journey made years before. With eyes closed they can draw accurate maps of their coastline. And their best maps drawn long ago rival the best maps we can make with satellite data. Their homing sense resembles that of animals that can find their way home through adverse conditions. They make little distinction between space and time. They observe details with keenness lost to Western people. Can design ingenuity be bred into a culture by adversity?

Such examples reveal the importance of the relation between culture and the long-term human prospect in particular places. There are, of course, many other examples, such as Helena Norberg-Hodge's (1992) study of the impact of Westernization on the people of Ladakh and Gary Nabhan's (1982) study of the Papago peoples of the desert Southwest. The history of settled people in many places reveals the fact that culture and the ecology of particular places have often been joined together with great intelligence and skill. The results, however imperfect, are habitats in which culture and nature have flourished together over many generations. They offer clues about how the human enterprise has, under some conditions, been sustained and what might be required to extend the life of our own.

Having been shaped by a century or more of cheap oil, industrialism, and hyperindividualism, we have a difficult time understanding what might be learned from such seemingly archaic examples. Yet as tourists we are drawn in large numbers to places like Amish country or Devon to snap a few photographs and after a brief visit return to other places that are not nearly as wholesome and to lives far more hectic. We seldom see any relation between the two. What can be learned from well-used landscapes and settled societies wherever they exist is the importance of local culture as the mediator between human intentions and nature. Design for settled peoples is more than the work of a few heroic individuals. The process by which cultures and communities evolve over long periods of time in particular places is manifest not so much in discrete and spectacular things as it is in overall stability and long-term prosperity. Indeed, it is the absence of spectacular monuments like pyramids, glittering office towers, and shopping malls that signals the intention of people to settle in and stay a while. Design in such places is a cultural process extending over many centuries that has certain identifiable characteristics.

In contrast to the frenetic pace of industrial societies, settled cultures work slowly, rather like "a patient and increasingly skillful lovemaking that [persuades] the land to flourish" (Hawkes 1951, 202). Moreover, settled cultures seldom exceed what can be called a human scale. They persist mostly, but not exclusively, on local resources. In Devon, most houses and barns are made from local timber and stone and roofed with local slate or thatch. Fences are grown as hedgerows over centuries. In Amish country, barns and houses are still built from local timber by the community in barn raisings. The culture is mostly powered by sunshine in the form of grass for animals and by wind for pumping water. Settled cultures grow most of their food. They provide their own livelihood. To their young they impart the skills and aptitudes necessary to live in a particular place, not the generic job skills necessary for the anywhere-and-everywhere industrial economy. Instead of individual brilliance, design results from an intelligence that is deeply embedded in the culture.

Settled cultures tend to limit excess in a variety of ways. Showiness, ego trips, great wealth, huge homes, hurry, and excessive consumption are mostly discouraged, while cooperation, neighborliness, competence, thrift, responsibility, and self-reliance are encouraged. I doubt that these traits are mentioned often, but they are manifest in the routines of daily life. It is simply the way things are. Western culture with its worship of egoism, doing your own thing, consumption, the cult of wealth, and keeping one's options open is simply incomprehensible from the viewpoint of settled people. Whatever their particular theology, settled cultures limit the expression of the seven deadly sins of pride, envy, anger, sloth, avarice, gluttony, and lust simply because these vices make living in close quarters difficult if not impossible. In Western culture, as Lewis Mumford (1961, 346) once noted, the deadly sins have mutated into "virtues" that feed economic obesity. When the two cultures have clashed, settled people have regarded industrial people as seriously deranged. But more often than not settled people are either subsequently seduced by materialism or swept away by the sheer power of the more aggressive culture.

→ Amish teenagers

Settled cultures, without using the word "ecology," have designed with ecology in mind because to do otherwise would bring ruin, famine, and social disintegration. Out of necessity they created harmony between intentions and the genius of particular places that preserved diversity both cultural and biological capital; utilized current solar income; created little or no waste; imposed few unaccounted costs; and supported cultural and social patterns. Cultures capable of doing such things work slowly and from the bottom up. There is no amount of individual cleverness that could have created the intricate cultural patterns that have preserved the landscape of Devon or grown rice in Bali for millennia, nor any that could have created a culture as stable and nondestructive as that of the Amish. On the contrary, these evolved as a continual negotiation within a community and between the community and the ecological realities of particular places. Such cultures are not the result of scientific research so much as continual trial and error at a scale small enough to give quick feedback on cause and effect. Ecological design, then, requires not just a set of generic design skills but rather the collective intelligence of a community of people applied to particular problems in a particular place over a long period of time.

Ecological design at the level of culture resembles the structure and behavior of resilient systems in other contexts in which feedback between action and subsequent correction is rapid, people are held accountable for their actions, functional redundancy is high, and control is decentralized. At a local scale, people's actions are known and so accountability tends to be high. Production is distributed throughout the community, which means that no one individual's misfortune disrupts the whole. Employment, food, fuel, and recreation are mostly derived locally, which means that people are buffered somewhat from economic forces beyond their control. Similarly, the decentralization of control to the community scale means that the pathologies of large-scale administration are mostly absent. Moreover, being situated in a place for generations provides long memory of the place and hence of its ecological possibilities and limits. There is a kind of long-term learning process that grows from the intimate experience of a place over time of the kind once described by English wheelwright George Sturt ([1923] 1984, 66) as "the age-long effort of Englishmen to fit themselves close and ever closer into England."

Beneath what we can see in settled cultures, there is a deeper worldview that we can barely comprehend. In contrast to the linear thinking characteristic of Westernized people, Native American cultures, for example, had a more integrated view of the world in which they lived. In Vine Deloria's words, "The traditional Indian stood in the center of a circle and brought everything together in that circle. Today we stand at the end of a line and work our way along that line, discarding or avoiding everything on either side of us" (1999, 257). There was (and for some, still is) a view that all that exists is bound in a kind of supportive kinship. These relationships imposed responsibilities on humans to perform tasks that upheld the "basic structure of the universe" and ensured that all life forms were treated with respect and dignity (ibid., 131). Humans were intended to live "as relatives" with all animals and learn from them (ibid., 237). "Apart from participation in this network," Deloria says, "Indians believe a person simply does not exist" (ibid., 132).

The idea that humans are embedded in a network of obligation and are kin to all life explains why settled cultures often regarded economics as a kind of gift relationship. "In most Indian communities in the old days the most respected person was the one who gave freely of physical wealth, who showed a concern for the unfortunate, and who allowed weaker members of the community to rely on him/her" (Deloria 1999, 132). The essence of the economy is the simple and profoundly ecological idea that "the gift must always move" (Hyde 1983, 4). Tribal people often evolved complicated ceremonies, like the potlatch of the Native American tribes of the North Pacific, in which wealth was given away, destroyed, or discarded. Beneath such customs is an ecological view of the world that involves understanding "that what nature gives to us is influenced by what we give to nature" (Deloria 1999, 19). When wealth is no longer regarded as a gift to be passed from person to person, then and only then does scarcity appear.

Such relationships were not religious abstractions, but central to the way Native Americans related to the places in which they lived. They made no clear distinctions between themselves physically and the land in which they dwelled. Land contained the memory of past deeds and the spirits of their ancestors. Settled people have always known where they would be buried and with whom. "Our memory of land is a memory of ourselves and our deeds and experiences," in Deloria's words (1999, 253). We who regard land as a commodity to be bought and sold or as a resource can scarcely comprehend such a view. Our lack of comprehension is, in the view of tribal people, a mark of our adolescence and immaturity.

This book is not an argument to return to some mythic condition of ecological innocence. No such place ever existed. It begins, however, with an acknowledgment that we have important things to relearn about the arts of longevity—what is now called "sustainability"—from earlier cultures and other societies. Many of those cultures appear to us as quaintly archaic if not utterly incomprehensible. But in the larger sweep of time, our emphasis on economic growth, consumption, and individualism will be even less understandable to subsequent and, one hopes, wiser generations. Carrying out the Great Work of making an ecologically durable and decent society will require us to confront the deeper cultural roots of our problems and grow out of the faith that we can meet the challenge of sustainability without really changing much. The evidence, I think, shows that we will have to change a great deal and mostly in ways that we will come to regard as vastly better than what exists now and certainly better than what is in prospect.

This is a design challenge like no other. It is not about making greener widgets but how to make decent communities that fit their places with elegant frugality. The issue is whether the emerging field of ecological design will evolve as a set of design skills applied as patchwork solutions on a larger pattern of disorder or whether design will eventually help to transform the larger culture that is badly in need of a reformation. I hope for the latter. Green consumerism or even greener corporations are Band-Aids on wounds inflicted by economy grown too indifferent to real human needs and pressing problems of long-term human survival. Corporations certainly need to be improved, but the larger design problem has to do with the structure of an economy that promotes excess consumption and human incompetence, concentrates power in too few hands, and destroys the ties that bind people together in community. The problem is not how to produce ecologically benign products for the consumer economy, but how to make decent communities in which people grow to be responsible citizens and whole people.

The essays that follow aim to broaden the concept of ecological design, explore various pathologies that prevent it, and sketch the educational implications of design. In the final section the essays lay out a standard for design that is oriented to generosity in the large sense of the word, the preservation of wildness and wilderness, and the design of a culture that protects its children.

2

Human Ecology as a Problem of Ecological Design

Man is everywhere a disturbing agent. Wherever he plants his foot, the harmonies of nature are turned to discords.

—*George Perkins Marsh*

The Problem of Human Ecology

Whatever their particular causes, environmental problems all share one fundamental trait: with rare exceptions they are unintended, unforeseen, and sometimes ironic side effects of actions arising from other intentions.[1] We intend one thing and sooner or later get something very different. We intended merely to be prosperous and

1. Our ecological troubles have been variously attributed to Judeo-Christian religion (White 1967), our inability to manage common property resources

healthy but have inadvertently triggered a mass extinction of other species, spread pollution throughout the world, and triggered climatic change—all of which undermines our prosperity and health. Environmental problems, then, are mostly the result of a miscalibration between human intentions and ecological results, which is to say that they are a kind of design failure.

The possibility that ecological problems are design failures is perhaps bad news because it may signal inherent flaws in our perceptual and mental abilities. On the other hand, it may be good news. If our problems are, to a great extent, the result of design failures, the obvious solution is better design, by which I mean a closer fit between human intentions and the ecological systems where the results of our intentions are ultimately played out.

The perennial problem of human ecology is how different cultures provision themselves with food, shelter, energy, and the means of livelihood by extracting energy and materials from their surroundings (Smil 1994). Ecological design describes the ensemble of technologies and strategies by which societies use the natural world to construct culture and meet their needs. Because the natural world is continually modified by human actions, culture and ecology are shifting parts of an equation that can never be solved. Nor can there be one correct design strategy. Hunter-gatherers lived on current solar income. Feudal barons extracted wealth from sunlight by exploiting serfs who farmed the land. We provision ourselves by mining ancient sunlight stored as fossil fuels. The choice is not whether or not human societies have a design strategy, but whether that strategy works ecologically and can be sustained within the regenerative capacity of the particular ecosystem. The problem of ecological design has become more difficult as the human population has grown and technology has multiplied. It is now the overriding problem of our time, affecting virtually all other issues on the human agenda. How and how intelligently we weave the human presence into the natural world will re-

---

such as ocean fisheries (Hardin 1968), lack of character (Berry 1977), gender imbalance (Merchant 1980), technology run amuck (Mumford 1974), disenchantment (Berman 1989), the loss of sensual connection to nature (Abram 1996), exponential growth (Meadows 1998), and flaws in the economic system (Daly 1996).

duce or intensify other problems having to do with ethnic conflicts, economics, hunger, political stability, health, and human happiness.

At the most basic level, humans need 2,200–3,000 calories per day, depending on body size and activity level. Early hunter-gatherers used little more energy than they required for food. The invention of agriculture increased the efficiency with which we captured sunlight permitting the growth of cities (Smil 1991, 1994). Despite their differences, neither hunter-gatherers nor farmers showed much ecological foresight. Hunter-gatherers drove many species to extinction, and early farmers left behind a legacy of deforestation, soil erosion, and land degradation. In other words, we have always modified our environments to one degree or another, but the level of ecological damage has increased with the level of civilization and with the scale and kind of technology.

The average citizen of the United States now uses some 186,000 calories of energy each day, most of it derived from oil and coal (Mc-Kibben 1998). Our food and materials come to us via a system that spans the world and whose consequences are mostly concealed from us. On average food is said to have traveled more than 1,300 miles from where it was grown or produced to where it is eaten (Meadows 1998). In such a system, there is no conceivable way that we can know the human or ecological consequences of eating. Nor can we know the full cost of virtually anything that we purchase or discard. We do know, however, that the level of environmental destruction has risen with the volume of stuff consumed and with the distance it is transported. By one count we waste more than 1 million pounds of materials per person per year. For every 100 pounds of product, we create 3,200 pounds of waste (Hawken 1997, 44). Measured as an "ecological footprint" (i.e., the land required to grow our food, process our organic wastes, sequester our carbon dioxide, and provide our material needs), the average North American requires some 5 hectares of arable land per person per year (Wackernagel and Rees 1996). But at the current population level, the world has only 1.2 hectares of useable land per person. Extending our lifestyle to everyone would require the equivalent of two additional earths!

Looking ahead, we face an imminent collision between a growing population with rising material expectations and ecological capacity. At some time in the next century, given present trends, the human population will reach or exceed 10 billion, perhaps as many as 15–20 percent of the species on earth will have disappeared forever, and the effects of climatic change will be fully apparent. This much and more is virtually certain. Feeding, housing, clothing, and educating another 4–6 billion people and providing employment for an additional 2–4 billion without wrecking the planet in the process will be a considerable challenge. Given our inability to meet basic needs of one-third of the present population, there are good reasons to doubt that we will be able to do better with the far larger population now in prospect.

The Default Setting

The regnant faith holds that science and technology will find a way to meet human needs and desires without our having to make significant changes in our philosophies, politics, economics, or in the way we live. Rockefeller University professor Jessie Ausubel, for example, asserts that

> after a very long preparation, our science and technology are ready also to reconcile our economy and the environment. . . . In fact, long before environmental policy became conscious of itself, the system had set decarbonization in motion. A highly efficient hydrogen economy, landless agriculture, industrial ecosystems in which waste virtually disappears: over the coming century these can enable large, prosperous human populations to co-exist with the whales and the lions and the eagles and all that underlie them. (Ausubel 1996, 15)

We have, Ausubel states, "liberated ourselves from the environment." This view is similar to that of futurist Herman Kahn when he asserted several decades ago that by the year 2200 "humans would everywhere be rich, numerous, and in control of the forces of nature" (Kahn and Brown 1976, 1). In its more recent version, those believing that we have liberated ourselves from the environment cite advances in energy use, materials science, genetic engineering, and artificial intelligence that will enable us to do much more with far less and eventually transcend ecological limits altogether. Humanity will then take control of its own fate, or more accurately, as C. S. Lewis once observed, some few humans will do so, purportedly acting on behalf of all humanity ([1947] 1970).

Ausubel's optimism coincides with the widely held view that we ought to simply take over the task of managing the planet (*Scientific American* 1989). In fact, the technological and scientific capability is widely believed to be emerging in the technologies of remote sensing, geographic information systems, computers, the science of ecology (in its managerial version), and systems engineering. The problems of managing the earth, however, are legion. For one, the word "management" does not quite capture the essence of the thing being proposed. We can manage, say, a 747 because we made it. Presumably, we know what it can and cannot do even though they sometimes crash for reasons that elude us. Our knowledge of the earth is in no way comparable. We did not make it, we have no blueprint of it, and we will never know fully how it works. Second, the target of management is not quite what it appears to be since a good bit of what passes for managing the earth is, in fact, managing human behavior. Third, under the guise of objective neutrality and under the pretext of emergency, management of the earth is ultimately an extension of the effort to dominate people through the domination of nature. And can we trust those presuming to manage to do so with fairness, wisdom, foresight, and humility, and for how long?

Another, and more modest, possibility is to restrict our access to nature rather like a fussy mother in bygone days keeping unruly children out of the formal parlor. To this end Martin Lewis (1992) proposes what he calls a "Promethean environmentalism" that aims to protect nature by keeping us away from as much of it as possible. His purpose is to substitute advanced technology for nature. This requires the development of far more advanced technologies, more unfettered capitalism, and probably some kind of high-tech virtual simulation to meet whatever residual needs for nature that we might retain in this Brave New World. Lewis dismisses the possibility that we could become stewards, ecologically competent, or even just a bit more humble. Accordingly, he disparages those whom he labels "eco-radicals," including Aldo Leopold, Herman Daly, and E. F. Schumacher, who question the role of capitalism in environmental destruction, raise issues about appropriate scale, and disagree with the directions of technological evolution. Lewis's proposal to protect nature by removing humankind from it raises other questions. Will people cut off from nature be sane? Will people who no longer believe that they need nature be willing, nonetheless, to protect it? If so, will people no longer in contact with nature know how to do so? And was it not our efforts to cut ourselves off from nature that got us into trouble in the first place? On such matters Lewis is silent.

Despite pervasive optimism, there is a venerable tradition of unease about the consequences of unconstrained technological development, from Mary Shelley's *Frankenstein* to Lewis Mumford's (1974) critique of the "megamachine." But the technological juggernaut that has brought us to our present situation, nonetheless, remains on track. We have now arrived, in Edward O. Wilson's (1998) view, at a choice between two very different paths of human evolution. One choice would aim to preserve "the physical and biotic environment that cradled the human species" along with those traits that make us distinctively human. The other path, based on the belief that we are now exempt from the "iron laws of ecology that bind other species," would take us in radically different directions, as "*Homo proteus* or 'shape-changer man'" (ibid., 278). But how much of the earth can we safely alter? How much of our own genetic inheritance should we manipulate before we are no longer recognizably human? This second path, in Wilson's view, would "render everything fragile" (ibid., 298). And, in time, fragile things break apart.

The sociologist and theologian Jacques Ellul, is even more pessimistic. "Our machines," he writes, "have truly replaced us." We have no philosophy of technology, in his view, because "philosophy implies limits and definitions and defined areas that technique will not allow" (1990, 216). Consequently, we seldom ask where all of this is going, or why, or who really benefits. The "unicity of the [technological] system," Ellul believes, "may be the cause of its fragility" (1980, 164). We are "shut up, blocked, and chained by the inevitability of the technical system," at least until the self-contradictions of the "technological bluff," like massive geologic fault lines, give way and the system dissolves in "enormous global disorder" (1990, 411–412). At that point he thinks that we will finally understand that "everything depends on the qualities of individuals" (ibid., 412).

The dynamic is by now familiar. Technology begets more technology, technological systems, technology-driven politics, technology-dependent economies, and finally, people who can neither function nor think a hair's breadth beyond the limits of one machine or another. This, in Neil Postman's (1992) view, is the underlying pattern of Western history as we moved from simple tools, to technocracy, to "technopoly." In the first stage, tools were useful to solve specific problems but did not undermine "the dignity and integrity of the culture into which they were introduced" (ibid., 23). In a technocracy like England in the eighteenth and nineteenth centuries, factories undermined "tradition, social mores, myth, politics, ritual and religion." The third stage, technopoly, however, "eliminates alternatives to itself in precisely the way Aldous Huxley outlined in *Brave New World*." It does so "by redefining what we mean by religion, by art, by family, by politics, by history, by truth, by privacy, by intelligence, so that our definitions fit its new requirements" (ibid., 48). Technopoly represents, in Postman's view, the cultural equivalent of AIDS, which is to say a culture with no defense whatsoever against technology or the claims of expertise (ibid., 63). It flourishes when the "tie between information and human purpose has been severed" (ibid., 70).

The course that Ausubel and others propose fits into this larger pattern of technopoly that step by step is shifting human evolution in radically different directions. Ausubel (1996) does not discuss the risks and unforeseen consequences that accompany unfettered technological change. These, he apparently believes, are justifiable as unavoidable costs of what he deems to be progress. This is precisely the kind of thinking that has undermined our capacity to refuse technologies that add nothing to our quality of life. A system that produces automobiles and atom bombs will also go on to make supercomputers, smart weapons, genetically altered crops, nanotechnologies, and eventually machines smart enough to displace their creators. There is no obvious stopping point, which is to say that, having accepted the initial premises of technopoly, the powers of control and good judgment are eroded away in the flood of possibilities.

Advertised as the essence of rationality and control, the technological system has become the epitome of irrationality in which means overrule careful consideration of ends. A rising tide of unanticipated consequences and "normal accidents" mock the idea that experts are in control or that technologies do only what they are intended to do. The purported rationality of each particular component in what Wilson (1998, 289) calls a "thickening web of prosthetic devices" added together as a system lacks both rationality and coherence. Nor is there anything inherently human or even rational about words such as "efficiency," "productivity," or "management," that are used to justify technological change. Rationality of this narrow sort has been "as successful—if not more successful—at creating new degrees of barbarism and violence as it has been at imposing reasonable actions" (Saul 1993, 32). Originating with Descartes and Galileo, the foundations of the modern worldview were flawed from the beginning. In time, those seemingly small and trivial errors of perception, logic, and heart cascaded into a rising tide of cultural incoherence, barbarism, and ecological degradation. Ausubel's optimism notwithstanding, this tide will continue to rise until it has finally drowned every decent possibility that might have been unless we choose a more discerning course.

Ecological Design

The unfolding problems of human ecology are not solvable by repeating old mistakes in new and more sophisticated and powerful ways. We need a deeper change of the kind Albert Einstein had in mind when he said that the same manner of thought that created problems could not solve them (quoted in McDonough and Braungart 1998, 92). We need what architect Sim van der Ryn and mathematician Stewart Cowan define as an ecological design revolution. Ecological design in their words is "any form of design that minimize(s) environmentally destructive impacts by integrating itself with living processes . . . the effective adaptation to and integration with nature's processes" (1996, x, 18). For landscape architect Carol Franklin, ecological design is a "fundamental revision of thinking and operation" (1997, 264). Good design does not begin with what we can do, but rather with questions about what we really want to do (Wann 1996, 22). Ecological design, in other words, is the careful meshing of human purposes with the larger patterns and flows of the natural world and the study of those patterns and flows to inform human actions (Orr 1994, 104).

In their book *Natural Capitalism* (1999), Paul Hawken, Hunter Lovins, and Amory Lovins propose a transformation in energy and resource efficiency that would dramatically increase wealth while using a fraction of the resources we currently use. Transformation would not occur, however, simply as an extrapolation of existing technological trends. They propose, instead, a deeper revolution in our thinking about the uses of technology so that we don't end up with "extremely efficient factories making napalm and throwaway beer cans" (Benyus 1997, 262). In contrast to Ausubel, the authors of *Natural Capitalism* propose a closer calibration between means and ends. Such a world would improve energy and resource efficiency by perhaps tenfold. It would be powered by highly efficient, small-scale, renewable energy technologies distributed close to the point of end-use. It would protect natural capital in the form of soils, forests, grasslands, oceanic fisheries, and biota while preserving biological diversity. Pollution, in any form, would be curtailed and eventually eliminated by industries designed to discharge no waste. The economy of that world would be calibrated to fit ecological realities. Taxes would be levied on things we do not want such as pollution and removed from things such as income and employment that we do want. These changes signal a revolution in design that draws on fields as diverse as ecology, systems dynamics, energetics, sustainable agriculture, industrial ecology, architecture, landscape architecture, and economics.[2]

The challenge of ecological design is more than simply an engineering problem of improving efficiency; it is the problem of reducing the rates at which we poison ourselves and damage the world. The revolution that van der Ryn and Cowan (1996) propose must first reduce the rate at which things get worse (coefficients of change) but eventually change the structure of the larger system. As Bill McDonough and Michael Braungart (1998) argue, we will need a second industrial revolution that eliminates the very concept of waste. This implies, as McDonough is fond of saying, "putting filters on our minds, not at the end of pipes." In practice, the change McDonough proposes

---

2. The roots of ecological design can be traced back to the work of Scottish biologist D'Arcy Thompson and his magisterial *On Growth and Form*, first published in 1917. In contrast to Darwin's evolutionary biology, Thompson traced the evolution of life forms back to the problems that elementary physical forces such as gravity pose for individual species. His legacy is an evolving science of forms evident in evolutionary biology, biomechanics, and architecture. Ecological design is evident in the work of Bill Browning, Herman Daly, Paul Hawken, Wes Jackson, Aldo Leopold, Amory and Hunter Lovins, John Lyle, Bill McDonough, Donella Meadows, Eugene Odum, Sim van der Ryn, and David Wann.

implies, among other things, changing manufacturing systems to eliminate the use of toxic and cancer-causing materials and developing closed-loop systems that deliver "products of service," not products that are eventually discarded to air, water, and landfills.

The pioneers in ecological design begin with the observation that nature has been developing successful strategies for living on earth for 3.8 billion years and is, accordingly, a model for

- Farms that work like forests and prairies
- Buildings that accrue natural capital like trees
- Waste water systems that work like natural wetlands
- Materials that mimic the ingenuity of plants and animals
- Industries that work more like ecosystems
- Products that become part of cycles resembling natural materials flows.

Wes Jackson (1985), for example, is attempting to redesign agriculture in the Great Plains to mimic the prairie that once existed there. Paul Hawken (1993) proposes to remake commerce in the image of natural systems. The new field of industrial ecology is similarly attempting to redesign manufacturing to reflect the way ecosystems work. The new field of "biomimicry" is beginning to transform industrial chemistry, medicine, and communications. Common spiders, for example, make silk that is ounce for ounce five times stronger than steel, with no waste by-products. The inner shell of an abalone is far tougher than our best ceramics (Benyus 1997, 97). By such standards, human industry is remarkably clumsy, inefficient, and destructive. Running through each of these ideas is the belief that the successful design strategies, tested over the course of evolution, provide the standard to inform the design of commerce and the large systems that supply us with food, energy, water, and materials, and remove our wastes (Benyus 1997).

The greatest impediment to an ecological design revolution is not, however, technological or scientific, but rather human. If intention is the first signal of design, as McDonough puts it, we must reckon with the fact that human intentions have been warped in recent history by violence and the systematic cultivation of greed, self-preoccupation, and mass consumerism. A real design revolution will have to transform human intentions and the larger political, economic, and institutional structure that permitted ecological degradation in the first place. A second impediment to an ecological design revolution is simply the scale of change required in the next few decades. All nations, but starting with the wealthiest, will have to:

- Improve energy efficiency by a factor of 5–10
- Rapidly develop renewable sources of energy
- Reduce the amount of materials per unit of output by a factor of 5–10
- Preserve biological diversity now being lost everywhere
- Restore degraded ecosystems
- Redesign transportation systems and urban areas
- Institute sustainable practices of agriculture and forestry
- Reduce population growth and eventually total population levels
- Redistribute resources fairly within and between generations
- Develop more accurate indicators of prosperity, well-being, health, and security.

To avoid catastrophe, all of these steps must be well under way within the next few decades. Given the scale and extent of the changes required, this is a transition for which there is no historical precedent. The century ahead will test, not just our ingenuity, but our foresight, wisdom, and sense of humanity as well.

The success of ecological design will depend on our ability to cultivate a deeper sense of connection and obligation without which few people will be willing to make even obvious and rational changes in time to make much difference. We will have to reckon with the power of denial, both individual and collective, to block change. We must reckon with the fact that we will never be intelligent enough to understand the full consequences of our actions, some of which will be paradoxical and some evil. We must learn how to avoid creating problems for which there is no good solution, technological or otherwise (Dobb 1996; Hunter 1997) such as the creation of long-lived wastes, the loss of species, or toxic waste flowing from tens of thousands of mines. In short, a real design revolution must aim to foster a deeper transformation in human intentions and the political and economic institutions that turn intentions into ecological results. There is no clever shortcut, no end-run around natural constraints, no magic bullet, and no such thing as cheap grace.

The Intention to Design

Designing a civilization that can be sustained ecologically and one that sustains the best in the human spirit will require us to confront the wellsprings of intention, which is to say, human nature. Our intentions are the product of many factors, at least four of which have implications for our ecological prospects. First, with the certain awareness of our mortality, we are inescapably religious creatures. The religious impulse in us works like water flowing up from an artesian spring that will come to the surface in one place or another. Our choice is not whether we are religious or not as atheists would have it, but whether the object of our worship is authentic or not. The gravity mass of our nature tugs us to create or discover systems of meaning that places us in some larger framework that explains, consoles, offers grounds for hope, and, sometimes, rationalizes. In our age, nationalism, capitalism, communism, fascism, consumerism, cyberism, and even ecologism have become substitutes for genuine religion. But whatever the -ism or the belief, in one way or another we will create or discover systems of thought and behavior that give us a sense of meaning and belonging to something larger. Moreover, there is good evidence to support the claim that successful resource management requires, in E. N. Anderson's words, "a direct, emotional religiously 'socialized' tie to the resources in question" (1996, 169). Paradoxically, however, societies with much less scientific information than we have often make better environmental choices. Myth and religious beliefs, which we regard as erroneous, have sometimes worked better to preserve environments than have decisions based on scientific information administered by presumably rational bureaucrats (Lansing 1991). Accordingly, solutions to environmental problems must be designed to resonate at deep emotional levels and be ecologically sound.

Second, despite all of our puffed up self-advertising as *Homo sapiens*, the fact is that we are limited, if clever, creatures. Accordingly, we need a more sober view of our possibilities. Real wisdom is rare and rarer still if measured ecologically. Seldom do we foresee the ecological consequences of our actions. We have great difficulty understanding what Jay Forrester (1971) once called the "counterintuitive behavior of social systems." We are prone to overdo what worked in the past, with the result that many of our current problems stem from past success carried to an extreme. Enjoined to "be fruitful and multiply," we did as commanded. But at 6 billion and counting, it seems that we lack the gene for enough. We are prone to overestimate our abilities to get out of self-generated messes. We are, as someone put it, continually overrunning our headlights. Human history is in large measure a sorry catalog of war and malfeasance of one kind or another. Stupidity is probably as great a factor in human affairs as intelligence. All of which is to say that a more sober reading of human potentials suggests the need for a fail-safe approach to ecological design that does not overtax our collective intelligence, foresight, and goodness.

Third, quite possibly we have certain dispositions toward the environment that have been hardwired in us over the course of our evolution. E. O. Wilson, for example, suggests that we possess what he calls "biophilia," meaning an innate "urge to affiliate with other forms of life" (1984, 85). Biophilia may be evident in our preference for certain landscapes such as savannas and in the fact that we heal more quickly in the presence of sunlight, trees, and flowers than in biologically sterile, artificially lit, utilitarian settings. Emotionally damaged children, unable to establish close and loving relationships with people, sometimes can be reached by carefully supervised contact with animals. And after several million years of evolution, it would be surprising indeed were it otherwise. The affinity for life described by Wilson and others, does not, however, imply nature romanticism, but rather something like a core element in our nature that connects us to the nature in which we evolved and which nurtures and sustains us. Biophilia certainly does not mean that we are all disposed to like nature or that it cannot be corrupted into biophobia. But without intending to do so, we are creating a world in which we do not fit. The growing evidence supporting the biophilia hypothesis suggests that we fit better in environments that have more, not less, nature. We do better with sunlight, contact with animals, and in settings that include trees, flowers, flowing water, birds, and natural processes than in their absence. We are sensuous creatures who develop emotional attachment to particular landscapes. The implication is that we need to create communities and places that resonate with our evolutionary past and for which we have deep affection.

Fourth, for all of our considerable scientific advances, our knowledge of the earth is still minute relative to what we will need to know. Where are we? The short answer is that despite all of our science, no one knows for certain. We inhabit the third planet out from a fifth-rate star located in a backwater galaxy. We are the center of nothing obvious to our science. We do not know whether the earth is just dead matter or whether it is, in some respects, alive. Nor do we know how forgiving the ecosphere may be to human insults. Our knowledge of the flora and fauna of the earth and the ecological processes that link them is small relative to all that might be known. In some areas, in fact, knowledge is in retreat because it is no longer fashionable or profitable. Our practical knowledge of particular places is often considerably less than that of the native peoples we displaced. As a result, the average college graduate would flunk even a cursory test on local ecology, and stripped of technology most would quickly founder.

To complicate things further, the advance of human knowledge is inescapably ironic. Since the Enlightenment, the goal of our science has been a more rational ordering of human affairs in which cause and effect could be empirically determined and presumably controlled. But after a century of promiscuous chemistry, for example, who can say how the 100,000 chemicals in common use mix in the ecosphere or how they might be implicated in declining sperm counts, rising cancer rates, disappearing amphibians, or behavioral disorders? And having disrupted global biogeochemical cycles, no one can say with assurance what the larger climatic and ecological effects will be. Undaunted by our ignorance, we rush ahead to reengineer the fabric of life on earth. Maybe scientists will figure it all out. It is more probable, however, that we are encountering the outer limits of social-ecological complexity in which cause and effect are widely separated in space and time, and in a growing number of cases no one can say with certainty what causes what. Like the sorcerer's apprentice, every answer generated by science gives rise to a dozen more questions, and every technological solution gives rise to even more problems. Rapid technological change intended to rationalize human life tends to expand the domain of irrationality. At the end of the bloodiest century in history, the Enlightenment faith in human rationality seems overstated at best. But the design implication is not less rationality, but a more complete, humble, and ecologically solvent rationality that works over the long term.

Who are we? Conceived in the image of God? Perhaps. But for the time being the most that can be said with assurance is that, in an evolutionary perspective, humans are a precocious and unruly newcomer with a highly uncertain future. Where are we? Wherever it is, it is a world full of irony and paradox, veiled in mystery. And for those purporting to establish the human presence in the world in a manner that is ecologically sustainable and spiritually sustaining, the ancient idea that God (or the gods) mocks human intelligence should never be far from our thoughts.

Ecological Design Principles

As creatures more ignorant than knowledgeable, what principles can safely guide our actions over the long term? There is no operating manual for planet Earth, so we will have to write our own as a set of design principles. Ecological design, however, is not so much about how to make things as about how to make things that fit gracefully over long periods of time in a particular ecological, social, and cultural context. Industrial societies, in contrast, work under the conviction that "if brute force doesn't work, you're not using enough of it." But when humans have designed with ecology in mind, there is greater harmony between intentions and the particular places in which those intentions are played out that preserves diversity both cultural and biological; utilizes current solar income; creates little or no waste; accounts for all costs; and respects larger cultural and social patterns. Ecological design is not just a smarter way to do the same old things or a way to rationalize and sustain a rapacious, demoralizing, and unjust consumer culture. The problem is not how to produce ecologically benign products for the consumer economy, but how to make decent communities in which people grow to be responsible citizens and whole people who do not confuse what they have with who they are. The larger design challenge is to transform a wasteful society into one that meets human needs with elegant simplicity. Designing ecologically requires a revolution in our thinking that changes the kinds of questions we ask from how can we do the same old things more efficiently to deeper questions such as:

- Do we need it?
- Is it ethical?
- What impact does it have on the community?
- Is it safe to make and use?
- Is it fair?
- Can it be repaired or reused?
- What is the full cost over its expected lifetime?
- Is there a better way to do it?

The quality of design, in other words, is measured by the elegance with which we join means and worthy ends. In Wendell Berry's felicitous phrase, good design "solves for pattern," thereby preserving the larger patterns of place and culture and sometimes this means doing nothing at all (1981, 134–145). In the words of John Todd, the aim is "elegant solutions predicated on the uniqueness of place."[3] Ecological design, then, is not simply a more efficient way to accommodate desires; it is the improvement of desire and all of those things that affect what we desire.

Ecological design is as much about politics and power as it is about ecology. We have good reason to question the large-scale plans to remodel the planet that range from genetic engineering to attempts to reengineer the carbon cycle. Should a few be permitted to redesign the fabric of life on the earth? Should others be permitted to design machines smarter than we are that might someday find us to be an annoyance and discard us? Who should decide how much of nature should be remodeled, for whose convenience, and by what standards? In an age when everything seems possible, where are the citizens or spokespersons for other members of biotic community who will be affected? The answer is that they are now excluded. At the heart of the issue of design, then, are procedural questions that have to do with politics, representation, and fairness.

It follows that ecological design is not so much an individual art practiced by individual designers as it is an ongoing negotiation between a community and the ecology of particular places. Good design

3. The phrase by John Todd is from a personal communication; see also John and Nancy Todd, *From Eco-Cities to Living Machines: Principles of Ecological Design* (Berkeley: North Atlantic Books, 1994).

results in communities in which feedback between action and subsequent correction is rapid, people are held accountable for their actions, functional redundancy is high, and control is decentralized. In a well-designed community, people would know quickly what's happening, and if they don't like it, they know who can be held accountable and can change it. Such things are possible only where livelihood, food, fuel, and recreation are, to a great extent, derived locally; where people have control over their own economies; and where the pathologies of large-scale administration are minimal. Moreover, being situated in a place for generations provides long memory of the place and hence of its ecological possibilities and limits. There is a kind of long-term learning process that grows from the intimate experience of a place over time. Ecological design, then, is a large idea but is most applicable at a relatively modest scale. The reason is not that smallness or locality has any necessary virtue, but that human frailties limit what we are able to comprehend and foresee, as well as the scope and consistency of our affections. No amount of smartness or technology can dissolve any of these limits. The modern dilemma is that we find ourselves trapped between the growing cleverness of our science and technology and our seeming incapacity to act wisely.

The standard for ecological design is neither efficiency nor productivity but health, beginning with that of the soil and extending upward through plants, animals, and people. It is impossible to impair health at any level without affecting it at other levels. The etymology of the word "health" reveals its connection to other words such as healing, wholeness, and holy. Ecological design is an art by which we aim to restore and maintain the wholeness of the entire fabric of life increasingly fragmented by specialization, scientific reductionism, and bureaucratic division. We now have armies of specialists studying bits and pieces of the whole as if these were separable. In reality it is impossible to disconnect the threads that bind us into larger wholes up to that one great community of the ecosphere. The environment outside us is also inside us. We are connected to more things in more ways than we can ever count or comprehend. The act of designing ecologically begins with the awareness that we can never entirely fathom those connections. This means that humans must act cautiously and with a sense of our fallibility.

Ecological design is not reducible to a set of technical skills. It is anchored in the faith that the world is not random but purposeful and stitched together from top to bottom by a common set of rules. It is grounded in the belief that we are part of the larger order of things and that we have an ancient obligation to act harmoniously within those larger patterns. It grows from the awareness that we do not live by bread alone and that the effort to build a sustainable world must begin by designing one that first nourishes the human spirit.

Finally, the goal of ecological design is not a journey to some utopian destiny, but is rather more like a homecoming. Philosopher Suzanne Langer once described the problem in these words: "Most people have no home that is a symbol of their childhood, not even a definite memory of one place to serve that purpose. Many no longer know the language that was once their mother-tongue. All old symbols are gone. . . . The field of our unconscious symbolic orientation is suddenly plowed up by the tremendous changes in the external world and in the social order" ([1942] 1976, 292). In other words, we are lost and must now find our way home again. For all of our technological accomplishments, the twentieth century was the most brutal and destructive era in our short history. In the century ahead we must chart a different course that leads to restoration, healing, and wholeness. Ecological design is a kind of navigation aid to help us find our bearings again. And getting home means recasting the human presence in the world in a way that honors ecology, evolution, human dignity, spirit, and the human need for roots and connection.

Conclusion

Ecological design is far more than the application of instrumental reason and advanced technology applied to the problems of shoehorning billions more of us into an earth already bulging at the seams with people. Humankind, as Abraham Heschel once wrote, "will not perish for want of information; but only for want of appreciation . . . what we lack is not a will to believe but a will to wonder" ([1951] 1990, 37). The ultimate object of ecological design is not the things we make but rather the human mind and specifically its capacity for wonder and appreciation.

The capacity of the mind for wonder, however, has been diminished by the tacit design of the systems that provide us with food, energy, materials, shelter, health care, entertainment, and by those that remove our voluminous wastes from sight and mind. There is little in these industrial systems that fosters mindfulness or ecological competence, let alone a sense of wonder. On the contrary, these systems are designed to generate cash, which has itself become an object of wonder and reverence. It is widely supposed that formal education serves as some kind of antidote to this uniquely modern form of barbarism. But conventional education, at its best, merely dilutes the tidal wave of false and distracting information embedded in the infrastructure and processes of technopoly. However well intentioned, formal education cannot compete with the larger educational effects of highways, shopping malls, supermarkets, urban sprawl, factory farms, agribusiness, huge utilities, multinational corporations, and nonstop advertising that teaches dominance, power, speed, accumulation, and self-indulgent individualism. We may talk about how everything is ecologically connected, but the terrible simplifiers are working overtime to take it all apart.

If it is not to become simply a more efficient way to do the same old things, ecological design must become a kind of public pedagogy built into the structure of daily life. There is little sense in only selling greener products to a consumer whose mind is still pre-ecological. Sooner or later that person will find environmentalism inconvenient, or incomprehensible, or too costly and will opt out. The goal is to calibrate human behavior with ecology, which requires a public that understands ecological possibilities and limits. To that end we must begin to see our houses, buildings, farms, businesses, energy technologies, transportation, landscapes, and communities in much the same way that we regard classrooms. In fact, they instruct in more fundamental ways because they structure what we see, how we move, what we eat, our sense of time and space, how we relate to each other, our sense of security, and how we experience the particular places in which we live. More important, by their scale and power they structure how we think, often limiting our ability to imagine better alternatives.

When we design ecologically, we are instructed continually by the fabric of everyday life: pedagogy informs infrastructure, which in turn informs us. Growing food on local farms and gardens, for example, becomes a source of nourishment for the body and instruction in soils, plants, animals, and cycles of growth and decay (Donahue 1999). Renewable energy technologies become a source of energy as well as insight about the flows of energy in ecosystems. Ecologically designed communities become a way to teach about land use, landscapes, and human connections. Restoration of wildlife corridors and habitats instructs us in the ways of animals. In other words, ecological design becomes a way to expand our awareness of nature and our ecological competence.

Most important, when we design ecologically we break the addictive quality that permeates modern life. "We have," in the words of Bruce Wilshire, "encase(d) ourselves in controlled environments called building and cities. Strapped into machines, we speed from place to place whenever desired, typically knowing any particular place and its regenerative rhythms and prospects only slightly" (1998, 18). We have alienated ourselves from "nature that formed our needs over millions of years [which] means alienation within ourselves" (ibid.). Given our inability to satisfy our primal needs, we suffer what Wilshire calls a "deprivation of ecstasy" that stemmed from the 99 percent of our life as a species spent fully engaged with nature. Having cut ourselves off from the cycles of nature, we find ourselves strangers in an alien world of our own making. Our response has been to create distractions and addictive behaviors as junk food substitutes for the totality of body-spirit-mind nourishment we've lost and then to vigorously deny what we've done. Ecstasy deprivation, in other words, results in surrogate behaviors, mechanically repeated over and over again, otherwise known as addiction. This is a plausible, even brilliant, argument with the ring of truth to it.

Ecological design is the art that reconnects us as sensuous creatures evolved over millions of years to a beautiful world. That world does not need to be remade but rather revealed. To do that, we do not need research as much as the rediscovery of old and forgotten things. We do not need more economic growth as much as we need to relearn the ancient lesson of generosity, as trustees for a moment between those who preceded us and those who will follow. Our greatest needs have nothing to do with the possession of things but rather with heart, wisdom, thankfulness, and generosity of spirit. And these virtues are part of larger ecologies that embrace spirit, body, and mind—the beginning of design.

§ 2

PATHOLOGIES AND BARRIERS

3

Slow Knowledge

There is no hurry, there is no hurry whatever.

—Erwin Chargaff

It takes all the running you can do, to keep in the same place.

—Lewis Carroll

Between 1978 and 1984 the Asian Development Bank spent $24 million to improve agriculture on the island of Bali. The target for improvement was an ancient agricultural system organized around 173 village cooperatives linked by a network of temples operated by "water priests" working in service to the water goddess, Dewi Danu, a diety seldom included in the heavenly pantheon of development economists. Not surprisingly, the new plan called for large capital investment to build dams and canals and to purchase pesticides and fertilizers. The plan also included efforts to make idle resources, both the Balinese and their land, productive year-round. Old practices of fallowing were ended, along with community celebrations and rituals. The results were remarkable but inconvenient: yields declined, pests proliferated, and the ancient village society began to unravel. On later examination (Lansing 1991), it turns out that the priests' role in the religion of Agama Tirtha was that of ecological master planners, whose task it was to keep a finely tuned system operating productively. Western development experts dismantled a system that had worked well for more than a millennium and replaced it with something that did not work at all. The priests have reportedly resumed control.

The story is a parable for much of the history of the twentieth century, in which increasingly homogenized knowledge is acquired and used more rapidly and on a larger scale than ever before and often with disastrous and unforeseeable consequences. The twentieth century is the age of fast knowledge driven by rapid technological change and the rise of the global economy. This has undermined communities, cultures, and religions that once slowed the rate of change and filtered appropriate knowledge from the cacophony of new information.

The culture of fast knowledge rests on these assumptions:

- Only that which can be measured is true knowledge
- The more knowledge we have, the better
- Knowledge that lends itself to use is superior to that which is merely contemplative
- The scale of effects of applied knowledge is unimportant
- There are no significant distinctions between information and knowledge
- Wisdom is an undefinable, hence unimportant, category.
- There are no limits to our ability to assimilate growing mountains of information, and none to our ability to separate essential knowledge from that which is trivial or even dangerous
- We will be able to retrieve the right bit of knowledge at the right time and fit it into its proper social, ecological, ethical, and economic context
- We will not forget old knowledge, but if we do, the new will be better than the old
- Whatever mistakes and blunders occur along the way can be rectified by yet more knowledge

- The level of human ingenuity will remain high
- The acquisition of knowledge carries with it no obligation to see that it is responsibly used
- The generation of knowledge can be separated from its application
- All knowledge is general in nature, not specific to or limited by particular places, times, and circumstances.

Fast knowledge is now widely believed to represent the essence of human progress. Although many admit the problems caused by the accumulation of knowledge, most believe that we have little choice but to keep on. After all, it's just human nature to be inquisitive. Moreover, research on new weapons and new corporate products is justified on the grounds that if we don't do it, someone else will and so we must. And increasingly, fast knowledge is justified on purportedly humanitarian grounds that we must hurry the pace of research to meet the needs of a growing population.

Fast knowledge has a lot going for it. Because it is effective and powerful, it is reshaping education, communities, cultures, lifestyles, transportation, economies, weapons development, and politics. For those at the top of the information society it is also exhilarating, perhaps intoxicating, and, for the few at the very top, it is highly profitable.

The increasing velocity of knowledge is widely accepted as sure evidence of human mastery and progress. But many, if not most, of the ecological, economic, social, and psychological ailments that beset contemporary society can be attributed directly or indirectly to knowledge acquired and applied before we had time to think it through carefully. We rushed into the fossil fuel age only to discover problems of acid precipitation and climate change. We rushed to develop nuclear energy without the faintest idea of what to do with the radioactive wastes. Nuclear weapons were created before we had time to ponder their full implications. Knowledge of how to kill more efficiently is rushed from research to application without much question about its effects on the perceptions and behavior of others, on our own behavior, or about better and cheaper ways to achieve real security. Chlorinated fluorocarbons, along with a host of carcinogenic, mutagenic, and hormone-disrupting chemicals, too, are products of fast knowledge. High-input, energy-intensive agriculture is also a product of knowledge applied before much consideration was given to its full ecological and social costs. Economic growth, in large measure, is driven by fast knowledge, with results everywhere evident in environmental problems, social disintegration, unnecessary costs, and injustice.

Fast knowledge undermines long-term sustainability for two fundamental reasons. First, for all of the hype about the information age and the speed at which humans are purported to learn, the facts say that our *collective* learning rate is about what it has always been: rather slow. A half-century after their deaths, for example, we have scarcely begun to fathom the full meaning of Gandhi's ideas about nonviolence or that of Aldo Leopold's "land ethic." Nearly a century and a half after *The Origin of Species*, we are still struggling to comprehend the full implications of evolution. And several millennia after Moses, Jesus, and Buddha, we are about as spiritually inept as ever. The problem is that the rate at which we collectively learn and assimilate new ideas has little to do with the speed of our communications technology or with the volume of information available to us, but it has everything to do with human limitations and those of our social, economic, and political institutions. Indeed, the slowness of our learning—or at least of our willingness to change—may itself be an evolved adaptation; short circuiting this limitation reduces our fitness.

Even if humans were able to learn more rapidly, the application of fast knowledge generates complicated problems much faster than we can identify and respond to them. We simply cannot foresee all the ways complex natural systems will react to human-initiated changes, at their present scale, scope, and velocity. The organization of knowledge by a minute division of labor further limits our capacity to comprehend whole-system effects, especially when the creation of fast knowledge in one area creates problems elsewhere at a later time. Consequently, we are playing catch up, but falling farther and farther behind. Finally, for reasons once described by Thomas Kuhn (1962), fast knowledge creates power structures that hold at bay alternative paradigms and worldviews that might slow the speed of change to manageable rates. The result is that the system of fast knowledge creates social traps in which the benefits occur in the near term while the costs are deferred to others at a later time.

The fact is that the only knowledge we've ever been able to count on for consistently good effect over the long run is knowledge that has been acquired slowly through cultural maturation. Slow knowledge is knowledge shaped and calibrated to fit a particular ecological and cultural context. It does not imply lethargy, but rather thoroughness and patience. The aim of slow knowledge is resilience, harmony, and the preservation of patterns that connect. Evolution is the archetypal example of slow knowledge. Except for rare episodes of punctuated equilibrium, evolution seems to work by the slow trial-and-error testing of small changes. Nature seldom, if ever, bets it all on a single throw of the dice. Similarly, every human culture that has artfully adapted itself to the challenges and opportunities of a particular landscape has done so by the patient and painstaking accumulation of knowledge over many generations; an age-long effort to fit close and ever closer into a particular place. Unlike fast knowledge generated in universities, think-tanks, and corporations, slow knowledge occurs incrementally through the process of community learning motivated more by affection than by idle curiosity, greed, or ambition. The worldview inherent in slow knowledge rests on these beliefs:

- Wisdom, not cleverness, is the proper aim of all true learning
- The velocity of knowledge can be inversely related to the acquisition of wisdom
- The careless application of knowledge can destroy the conditions that permit knowledge of any kind to flourish (a nuclear war, for example, made possible by the study of physics, would be detrimental to the further study of physics)
- What ails us has less to do with the lack of knowledge but with too much irrelevant knowledge and the difficulty of assimilation, retrieval, and application as well as the lack of compassion and good judgment
- The rising volume of knowledge cannot compensate for a rising volume of errors caused by malfeasance and stupidity generated in large part by inappropriate knowledge
- The good character of knowledge creators is not irrelevant to the truth they intend to advance and its wider effects
- Human ignorance is not an entirely solvable problem; it is, rather, an inescapable part of the human condition.

The differences between fast knowledge and slow knowledge could not be more striking. Fast knowledge is focused on solving problems, usually by one technological fix or another; slow knowledge has to do with avoiding problems in the first place. Fast knowledge deals with discrete problems, whereas slow knowledge deals with context, patterns, and connections. Fast knowledge arises from hierarchy and competition; slow knowledge is freely shared within a community. Fast knowledge is about know-how; slow knowledge about is about know-how and know-*why*. Fast knowledge is about competitive edges and individual and organizational profit; slow knowledge is about community prosperity. Fast knowledge is mostly linear; slow knowledge is complex and ecological. Fast knowledge is characterized by power and instability; slow knowledge is known by its elegance, complexity, and resilience. Fast knowledge is often regarded as private property; slow knowledge is owned by no one. In the culture of fast knowledge, man is the measure of all things. Slow knowledge, in contrast, occurs as a co-evolutionary process among humans, other species, and a shared habitat. Fast knowledge is often abstract and theoretical, engaging only a portion of the mind. Slow knowledge, in contrast, engages all of the senses and the full range of our mental powers. Fast knowledge is always new; slow knowledge often is very old. The besetting sin inherent in fast knowledge is hubris, the belief in human omnipotence now evident on a global scale. The sin of slow knowledge can be parochialism and resistance to needed change.

Are there occasions when we need fast knowledge? Yes, but with the caveat that a significant percentage of the problems we now attempt to solve quickly through complex and increasingly expensive means have their origins in the prior applications of fast knowledge. Solutions to such problems often resemble a kind of Rube Goldberg contraption that produces complicated, expensive, and often temporary cures for otherwise unnecessary problems. The point, as every accountant knows, is that there is a difference between gross and net. And after all of the costs of fast knowledge are subtracted, the net gains in many fields have been considerably less than we have been led to believe.

What can be done? Until the sources of power that fuel fast knowledge run dry, perhaps nothing. Then again, maybe we are not quite so powerless as that. The problem is clear: we need no more fast knowledge cut off from its ecological and social context, which is ignorant knowledge. In principle, the solution is equally clear: we need to discover and sometimes rediscover the knowledge of things such as how the earth works, how to build sustainable and sustaining communities that fit their regions, how to raise and educate children to be decent people, and how to provision ourselves justly and within ecological limits. We need to remember all of those things necessary to re-member a world fractured by competition, fear, greed, and short-sightedness. If there is no quick cure, neither are we without the wherewithal to create a better balance between the real needs of society and the pace and kind of knowledge generated. For colleges and universities, in particular, I propose the following steps aimed to improve the quality of knowledge by slowing its acquisition to a more manageable rate.

First, scholars ought to be encouraged to include practitioners and those affected in setting priorities and standards for the acquisition of knowledge. Professionalized knowledge is increasingly isolated from the needs of real people and, to that extent, dangerous to our larger prospects. It makes no sense to rail about participation in the political and social affairs of the community and nation while allowing the purveyors of fast knowledge to determine the actual conditions in which we live without so much as a whimper. Knowledge has social, economic, political, and ecological consequences as surely as any act of Congress, and we ought to demand representation in the setting of research agendas for the same reason that we demand it in matters of taxation. Inclusiveness would slow research to more manageable rates while improving its quality. There are good examples of participatory research involving practitioners in agriculture (Hassanein 1999), forestry (Banuri and Marglin 1993), land use (Appalachian Land Ownership Task Force 1983), and urban policy. There should be many more.

Second, faculty ought to be encouraged in every way possible to take the time necessary to broaden their research and scholarship to include its ecological, ethical, and social context. They ought to be encouraged to rediscover old and true knowledge and to respect prior wisdom. And colleges and universities could do much more to encourage and reward efforts by their faculty to teach well and to apply existing knowledge to solve real problems in their communities.

Third, colleges and universities ought to foster a genuine and ongoing debate about the velocity of knowledge and its effects on our larger prospects. We bought in to the ideology that faster is better without taking the time to think it through. Increasingly, we communicate by electronic mail and the Internet. As a consequence, I believe that one can detect a decline in the salience of our communication and perhaps in its civility as well in direct proportion to its velocity and volume. It is certainly possible to detect a growing frustration among faculty with the time it takes to separate chaff from the grain in the rising deluge of e-mail, regular mail, memos, administrative pronouncements, and directives.

Conclusion

Fast knowledge has played havoc in the world because *Homo sapiens* is just not smart enough to manage everything that it is possible for the human mind to discover and create. In Wendell Berry's words, there is a kind of idiocy inherent in the belief "that we can first set demons at large, and then, somehow, become smart enough to control them" (1983, 65). Slow knowledge really isn't slow at all. It is knowledge acquired and applied as rapidly as humans can comprehend it and put it to consistently good use. Given the complexity of the world and the depth of our human frailties, this takes time and it always will. Mere information can be transmitted and used quickly, but new knowledge is something else. Often it requires rearranging worldviews and paradigms, which we can only do slowly. Instead of increasing the speed of our chatter, we need to learn to listen more attentively. Instead of increasing the volume of our communication, we ought to improve its content. Instead of communicating more extensively, we should converse more intensively with our neighbors without the help of any technology whatsoever. "There is no hurry, there is no hurry whatever."

4

Speed

But is the nature of civilization "speed"? Or is it "consideration"? Any animal can rush around a corral four times a day.
Only a human being can consciously oblige himself to go
slowly in order to consider whether he is doing the right thing,
doing it the right way, or ought in fact to be doing something
else. . . . Speed and efficiency are not in themselves signs of intelligence or capability or correctness.

—*John Ralston Saul*

Water

Plum Creek begins in drainage from farms on the west side of the city
of Oberlin, Ohio, and flows eastward through a city golf course, a
college arboretum, and the downtown area. East of the city, the
stream receives the effluent from the city sewer facility before it joins
with the Black River, which flows north through two rust-belt cities,
Elyria and Lorain, before emptying into Lake Erie 25 miles west of

Cleveland. Plum Creek shows all of the signs of 150 years of human use and abuse. As late as 1850 the stream ran clear even in times of flood, but now it is murky brown year-round. Because of pollution, sediments, and the lack of aquatic life, the U.S. Environmental Protection Agency considers it to be a "nonattainment" stream. Yet it survives, more or less. To most residents of Oberlin, Plum Creek is little more than a drain and sewer useful for moving water off the land as rapidly as possible. Few regard it as an aesthetic asset or ecological resource.

The character of Plum Creek changes quickly as it flows eastward into downtown Oberlin. Runoff from city streets enters the stream where the creek runs under the intersection of Morgan and Professor Streets. One block to the east, a larger volume of runoff polluted by oil and grease from city streets enters the creek as it flows under Main Street, past a Midas Muffler shop, a NAPA Auto Parts Store, and City Hall, located in the flood plain. Where Plum Creek flows under Main Street, an increased volume of storm water and consequently increased stream velocity have widened the banks and cut the channel from several feet to a depth of 10 feet or more. The city has attempted to stabilize the stream by lining the banks with concrete or by riprapping with large chunks of broken concrete. The aquatic life that exists upstream mostly disappears as Plum Creek flows through the downtown. Bending to the northeast, the creek passes through suburban backyards, past the municipal wastewater plant, a Browning Ferris Industries landfill, and on toward the west fork of the Black River and Lake Erie.

Whatever Plum Creek once was, it is now fundamentally shaped by the fact that European settlers cut the forests and drained marshes which once absorbed rainfall and released water slowly throughout the year. The wetlands and forests that once made up the flood plain are now mostly gone, replaced by roads, lawns, buildings, and parking lots. Rainfall is quickly channeled from lawns, streets, and parking lots into storm drains and culverts and diverted into the creek. The result is a landscape that sheds water quickly, contributing to floods, reducing water quality, and degrading aquatic habitats. Mathematics tells the story: doubling the speed of water increases the size of soil particles transported by 64 times.

The history of the Plum Creek watershed is not unusual. More than 90 percent of Ohio wetlands have been drained. As a nation, we have lost more than 50 percent of the wetlands that existed before European settlement and despite federal laws we continue to lose wetlands at a net rate of 24,000 acres each year (Revkin 2001, 1). The total paved area in the lower 48 states is equivalent to a land area larger than Kentucky. As a result, water moves more quickly across our landscapes than it once did, so that flooding, particularly downstream from urban areas, is more common and more severe than ever. Measured in constant dollars, flood plain damage rose by 50 percent between 1975 and 1990. We labor in vain to control flooding and prevent flood damage by the heroic engineering of dams, levees, and diversion channels while continuing to clear forests, drain wetlands, and pave. The results shown in the Mississippi floods of 1993 or those along the Missouri and Ohio rivers in 1997 are now part of the escalating price we pay for engineering, as if the velocity of water moving through the landscape did not matter.

Money

The city of Oberlin is a fairly typical midwestern college town with a square around which are arrayed college buildings, a historic church, an art museum, a hotel, and downtown businesses including three banks, two book stores, a bakery, a five and dime store, an Army-Navy store, an assortment of restaurants, a gourmet coffee shop, pizza parlors, and one struggling hardware store. In the past six years, however, the downtown lost among other businesses a car dealership, a drug store, a bicycle repair shop, and stores selling auto parts, clothing, and appliances. Going back even further, the economic changes are more striking. Older residents remember the six grocery stores that would deliver to your home, local dairies that delivered milk in glass bottles, and a train station. All that changed after World War II. A large mall with the standard assortment of national merchants located 10 miles away now drains off the largest part of what had once been mostly local business. Going south out of town, new development in Oberlin begins unsurprisingly with a McDonalds and a chain drug store. Farther on, a Pizza Hut newly relocated from the downtown has opened beside a large discount store with more strip development on the way. If this sounds familiar, it should. It is the American pattern of automobile-driven development by which capital moves from older downtowns to the periphery where land is cheaper and zoning regulations are more lax.

Despite the fact that the city includes a well-endowed college, a vocational school, an air traffic control center, and an industrial park, an estimated 38 percent of the residents of Oberlin live below the poverty line. Money does not stay in the local economy for long. Most of the salaries and wages paid out in Oberlin exit the city economy quickly. Hence the multiplier effect or the number of times a dollar is spent in the local economy before being used to purchase something outside is low.

In contrast, 55 miles to the south in the Amish economy of Holmes County, the economic multiplier would be very high and unemployment and poverty virtually nonexistent. The Amish buy and sell from each other. They make their own tools, farm implements, and furniture. They grow a large percentage of their food, much of which they process themselves so that the value is added locally. Their expenditures for fuel, health care, consumer goods, luxury items, and expensive items like cars or retirement costs are low to zero. They have their own insurance system, which to a great extent consists of the applied arts of neighborliness toward those in need. They accept neither welfare nor social security. The contrast between the Oberlin economy and that of the Amish could hardly be greater.

An Amish friend of mine recently told me that "the horse is the salvation of the Amish society." The Amish culture, as previously noted, operates at the speed of the horse and the sun. Because they farm with horses, they aren't tempted to farm large amounts of land. Farming with horses, in other words, serves as a brake to the temptation to take over a neighbor's land. And because the effective radius of a horse-drawn buggy is about eight miles, and its hauling capacity is low, the Amish are not much tempted by consumerism at the local mall. But horsespeed does more. It slows the velocity of work to a pace that allows close observation of soils, wildlife, and plants. My Amish friend often uses only a walking plow, which he believes preserves soil biota and prevents erosion. The speed of the horse, in other words, allows the Amish to pay attention to the minute particulars of their farm and how they farm. By a similar logic, he waits to cut hay until the bobolinks in the field have fledged. The loss in protein content in the hay he believes is more than compensated by the health of the place and the pleasure derived from having birds on the farm.

The capital tied up in an Amish farm is mostly in land and buildings, not in equipment. Their cash flow seldom goes to banks or vendors of petrochemicals and fossil fuels. It is small wonder that Amish farms continue to thrive while 4.5 million non-Amish farms have disappeared in the past 60 years.

Information

Several years ago the college where I teach created an electronic "quick mail" system to reduce paper use and to increase our efficiency. Electronic communication is now standard throughout most organizations. The results, however, are mixed at best. The most obvious result is a large increase in the sheer volume of stuff communicated, much of which is utterly trivial. There is also a manifest decline in the grammar, literary style, and civility of communication. People stroll down the hall or across campus to converse less frequently than before. Students remain transfixed before computer screens for hours, often doing no more than playing computer games. Our conversations, thought patterns, and institutional speed are increasingly shaped to fit the imperatives of technology. Not surprisingly, more and more people feel overloaded by the demands of incessant communication. But to say so publicly is to run afoul of the technological fundamentalism now dominant virtually everywhere.

By default and without much thought, it has been decided (or decided for us) that communication ought to be cheap, easy, and quick. Accordingly, more and more of us are instantly wired to the global nervous system with cell phones, beepers, pagers, fax machines, and e-mail. If useful in real emergencies, the overall result is to homogenize the important with the trivial, making everything an emergency and an already frenetic civilization even more frenetic. As a result, we are drowning in unassimilated information, most of which fits no meaningful picture of the world. In our public affairs and in our private lives we are, I think, increasingly muddle-headed because we have mistaken volume and speed of information for substance and clarity.

On my desk I have the three volumes of correspondence between Thomas Jefferson and James Madison written with quill pen by candlelight and delivered by horse. The style is mostly impeccable.

Even when they wrote about mundane things, they did so with clarity and insight. Their disagreements were expressed with civility and felicity. The entire body of letters can be read for both pleasure and instruction. Assuming people still read two centuries from now, will they read the correspondence of, say, Bill Clinton or George W. Bush for either pleasure or instruction? In contrast to our own, Jefferson and Madison were part of a culture that, whatever its other flaws, had time to take words seriously. They knew, intuitively perhaps, that information and knowledge were not the same thing and that neither was to be confused with wisdom. In large part the difference, whether they thought about it or not, was the speed of the society.

It is time to consider the possibility that, for the most part, communication ought to be somewhat slower, more difficult, and more expensive than it is now. Beyond some relatively low threshold, the rapid movement of information works against the emergence of knowledge, which requires the time to mull things over, to test results, and, when warranted, to change perceptions and behavior. The speed of genuine wisdom, which requires the integration of many different levels of knowledge, is slower still. Only over generations through a process of trial and error can knowledge eventually congeal into cultural wisdom about the art of living well within the resources, assets, and limits of a place.

Synthesis

Water moving too quickly through a landscape does not recharge underground aquifers. The results are floods in wet weather and droughts in the summer. Money moving too quickly through an economy does not recharge the local wellsprings of prosperity, whatever else it does for the global economy. The result is an economy polarized between those few who do well in a high-velocity economy and those left behind. Information moving too quickly to become knowledge and grow into wisdom does not recharge moral aquifers on which families, communities, and entire nations depend. The result is moral atrophy and public confusion. The common thread between all three is velocity. And they are tied together in a complex system of cause and effect that we have mostly overlooked.

There is an appropriate velocity for water set by geology, soils, vegetation, and ecological relationships in a given landscape. There is an appropriate velocity for money that corresponds to long-term needs of whole communities rooted in particular places and the necessity of preserving ecological capital. There is an appropriate velocity for information, set by the assimilative capacity of the mind and by the collective learning rate of communities and entire societies. Having exceeded the speed limits, we are vulnerable to ecological degradation, economic arrangements that are unjust and unsustainable, and, in the face of great and complex problems, to befuddlement that comes with information overload.

The ecological impacts of increased velocity of water are easy to comprehend. We can see floods, and with effort we can discern how human actions can amplify droughts. But it is harder to comprehend the social, political, economic, and ecological effects of increasing velocity of money and information, which are often indirect and hidden. Increasing velocity of commerce, information, and transport, however, requires more administration and regulation of human affairs to ameliorate congestion and other problems. More administration means that there are fewer productive people, higher overhead, and higher taxes to pay for more infrastructure necessitated by the speed of people and things and problems of congestion. Increasing velocity and scale tends to increase the complexity of social and ecological arrangements and reduce the time available to recognize and avoid problems. Cures for problems caused by increasing velocity often set in motion a cascading series of other problems. As a result, we stumble through a succession of escalating crises with diminishing capacity to act intelligently. Other examples fit the same pattern such as the velocity of transportation, material flows, extraction of nonrenewable resources, introduction of new chemicals, and human reproduction. At the local scale the effect is widening circles of disintegration and social disorder. At the global scale, the rate of change caused by increasing velocity disrupts biological evolution and the biogeochemical cycles of the earth.

The increasing velocity of the global culture is no accident. It is the foundation of the corporation-dominated global economy that requires quick returns on investment and the obsession with rapid economic growth. It is the soul of the consumer economy that feeds on impulse, obsession, and instant gratification. The velocity of water in our landscape is a direct result of too many automobiles, too much paving, sprawling development, deforestation, and a food system that cannot be sustained in any decent or safe manner. The speed of information is driven by something that more and more resembles addiction. But above all, increasing speed is driven by minds unaware of the irony that the race has never been to the swift.

Upshot

We are now engaged in a great global debate about how we might lengthen our tenure on the earth. The discussion is mostly confined to options having to do with better technology, more accurate resource prices, and smarter public policies, all of which are eminently sensible, but hardly sufficient. The problem is simply how a species pleased to call itself *Homo sapiens* fits on a planet with a biosphere. This is a design problem and requires a design philosophy that takes time, velocity, scale, evolution, and ecology seriously. We will neither conserve biotic resources nor build a sustainable civilization that operates at our present velocity.

But here's the rub: The very ideas that we need to build a sustainable civilization need to be invented or rediscovered, then widely disseminated, and put into practice quickly. Yet the same forces that have combined to give us a high-velocity economy and society reform themselves at glacial speed. Nearly 140 years after *The Origin of Species*, we still farm as if evolution did not matter. More than three decades after *Silent Spring*, we use more synthetic chemicals than ever. Three decades after publication of *The Limits to Growth* (Meadows et al. 1972), economic obesity is still the goal of governments everywhere. And a quarter of a century after Amory Lovins's prophetic and, as it turns out, understated projections about the potential for energy efficiency and solar energy (Lovins 1976), we are still using two to three times more fossil fuel than we need. Wendell Berry's devastating critique of American agriculture was published in 1977, yet sustainable agriculture is still a distant dream. Nearly a decade has passed since the scientific consensus began to form about the seriousness of global warming, yet we dawdle. I could go on, but the point is clear. The things that need to happen rapidly such as the preservation of biological diversity, the transition to a solar society, the widespread application of sustainable agriculture and forestry, population limits, the protection of basic human rights, and democratic reform occur slowly, if at all, while ecological ruin and economic dislocation race ahead. What can be done?

First, we need a relentless analytical clarity to discern the huge inefficiencies of high-speed "efficiency." We have contrived a high-technology, high-speed economy that is neither sustainable nor capable of sustaining what is best in human cultures. On close examination, many of the alleged benefits of ever-rising affluence are fraudulent claims. Thoughtful analysis reveals that our economy often works to do with great expense, complication, and waste things that could be done more simply, elegantly, and harmoniously or in some cases things that should not be done at all. Most of our mistakes were a result of hurry in the name of economic competition, or national security, or progress. Now many mistakes must be expensively undone or written off as a permanent loss. The speed of the industrial economy must be reset to take account of evolution, natural rhythms, and genuine human needs. That means recalibrating public policies and taxation to promote a more durable prosperity.

Next, we need a more robust idea of time and scale that takes the health of people and communities seriously:

> The only way that can induce us to reduce our speed of movements is a return to a spatially more contracted, leisurely, and largely pedestrian mode of life that makes high speeds not only unnecessary but as uneconomic as a Concorde would be for crossing the English Channel. . . . In other words, slow is beautiful in an appropriately contracted small social environment of beehive density and animation not only from a political and economic but, in the most literal sense, also from an aesthetic point of view, releasing an abundance of long abandoned energy not by patriotically making us drive slowly, but by depriving us materially of the need for driving fast. (Kohr 1980, 58)

Our assumptions about time are crystallized in community design and architecture. Sprawling cities, economic dependency, and long-distance transport of food and materials require high-velocity transport, high-speed communication, and result in higher costs, community disintegration, and ecological deformation. Rethinking velocity and time will require rethinking our relationship to the land as well. Here, too, we have options for increasing density through open space development and smarter planning that create proximity between housing, employment, shopping, culture, public spaces, recreation, and health care—what is now being called the "new urbanism."

Finally, in a society in which people sometimes talk about "killing time" we must learn, rather, to take time. We must learn to take time to study nature as the standard for much of what we need to do. We must take time and make the effort to preserve both cultural and biological diversity. We must take time to calculate the full costs of what we do. We must take time to make things durable, repairable, useful, and beautiful. We must take the time, not just to recycle, but rather to eliminate the very concept of waste. In most things, timeliness and regularity, not speed, are important. Genuine charity, good parenting, true neighborliness, good lives, decent communities, conviviality, democratic deliberation, real prosperity, mental health, and the exercise of true intelligence have a certain pace and rhythm that can only be harmed by being accelerated. The means to control velocity can be designed into daily life like speed bumps designed to slow auto traffic. Holidays, festivals, celebrations, sabbaticals, Sabbaths, prayer, good conversation, storytelling, music making, the practice of fallowing, shared meals, a high degree of self-reliance, craftwork, walking, and shared physical work are speed control devices used by every healthy culture.

5

Verbicide

In the beginning was the Word.

—*John 1:1*

He entered my office for advice as a freshman advisee sporting nearly perfect SAT scores and an impeccable academic record—by all accounts a young man of considerable promise. During a 20-minute conversation about his academic future, however, he displayed a vocabulary that consisted mostly of two words: "cool" and "really." Almost 800 SAT points hitched to each word. To be fair, he could use them interchangeably as "really cool" or "cool . . . really!" He could also use them singly, presumably for emphasis. When he became one of my students in a subsequent class I confirmed that my first impression of the young scholar was largely accurate and that his vocabulary, and presumably his mind, consisted predominantly of words and images derived from overexposure to television and the new jargon of computer-speak. He is no aberration, but an example of a larger problem, not of illiteracy but of diminished literacy in a culture that often sees little reason to use words carefully, however abundantly. Increasingly, student papers, from otherwise very good students, have whole paragraphs that sound like advertising copy. Whether students are talking or writing, a growing number have a tenuous grasp on a declining vocabulary. Excise "uh . . . like . . . uh" from virtually any teenage conversation, and the effect is like sticking a pin into a balloon.

In the past 50 years, by one reckoning, the working vocabulary of the average 14-year-old has declined from some 25,000 words to 10,000 words ("Harper's Index" 2000). This reflects not merely a decline in numbers of words but in the capacity to think. It also reflects a steep decline in the number of things that an adolescent needs to know and to name in order to get by in an increasingly homogenized and urbanized consumer society. This is a national tragedy virtually unnoticed in the media. It is no mere coincidence that in roughly the same half century the average person has learned to recognize more than 1,000 corporate logos but can recognize fewer than 10 plants and animals native to their locality (Hawken 1993, 214). That fact says a great deal about why the decline in working vocabulary has gone unnoticed—few are paying attention. The decline is surely not consistent across the full range of language but concentrates in those areas having to do with large issues such as philosophy, religion, public policy, and nature. On the other hand, vocabulary has probably increased in areas having to do with sex, violence, recreation, and consumption. As a result, we are losing the capacity to say what we really mean and ultimately to think about what we mean. We are losing the capacity for articulate intelligence about the things that matter most. "That sucks," for example, is a common way for budding young scholars to announce their displeasure about any number of issues that range across the full spectrum of human experience. But it can also be used to indicate a general displeasure with the entire cosmos. Whatever the target, it is the linguistic equivalent of using duct tape for holding disparate thoughts in rough proximity to some vague emotion of dislike.

The problem is not confined to teenagers or young adults. It is part of a national epidemic of incoherence evident in our public discourse, street talk, movies, television, and music. We have all heard popular music that consisted mostly of pre-Neanderthal grunts. We have witnessed "conversation" on TV talk shows that would have embarrassed retarded chimpanzees. We have listened to politicians of national reputation proudly mangle logic and language in less than a paragraph, although they can do it on a larger scale as well. However manifested, it is aided and abetted by academics, including whole departments specializing in various forms of postmodernism and the deconstruction of one thing or another. They propounded ideas that everything was relative, hence largely inconsequential, and that the use of language was an exercise in power, hence to be devalued. They taught, in other words, a pseudo-intellectual contempt for clarity, careful argument, and felicitous expression. Being scholars of their word, they also wrote without clarity, argument, and felicity. Remove half a dozen arcane words from any number of academic papers written in the past 10 years and the argument—whatever it was—evaporates. But the situation is not much better elsewhere in the academy where thought is often fenced in by disciplinary jargon. The fact is that educators have all too often been indifferent trustees of language. This explains, I think, why the academy has been a lame critic of what ails the world from the preoccupation with self to technology run amuck. We have been unable to speak out against the barbarism engulfing the larger culture because we are part of the process of barbarization that begins with the devaluation of language.

The decline of language, noted by commentators such as H. L. Mencken, George Orwell, William Safire, and Edwin Newman, is nothing new. Language is always coming undone. Why? For one thing, it is always under assault by those who intend to control others by first seizing the words and metaphors by which people describe their world. The goal is to give partisan aims the appearance of inevitability by diminishing the sense of larger possibilities. In our time language is under assault by those whose purpose it is to sell one kind of quackery or another: economic, political, religious, or technological. It is under attack because the clarity and felicity of language (as distinct from its quantity) is devalued in an industrial-technological society. The clear and artful use of language is, in fact, threatening to that society. As a result we have highly distorted and atrophied conversations about ultimate meanings, ethics, public purposes, or the means by which we live. Since we cannot solve problems that we cannot name, one result of our misuse of language is a growing agenda of unsolved problems that cannot be adequately described in words and metaphors derived from our own creations such as machines and computers.

Second, language is in decline because it is being balkanized around the specialized vocabularies characteristic of an increasingly specialized society. The highly technical language of the expert is, of course, both bane and blessing. It is useful for describing fragments of the world, but not for describing how these fit into a coherent whole. But things work as whole systems, whether or not we can say it and whether or not we perceive it. And more than anything else, it is coherence our culture lacks, not specialized knowledge. Genetic engineering, for example, can be described as a technical matter in the language of molecular biology. But saying what the act of rearranging the genetic fabric of earth means requires an altogether different language and a mind-set that seeks to discover larger patterns. Similarly, the specialized language of economics does not begin to describe the state of our well-being, whatever it reveals about how much we may or may not possess. Regardless of these arguments, over and over the language of the specialist trumps that of the generalist—the specialist in whole things. The result is that the capacity to think carefully about ends, as distinct from means, has all but disappeared from our public and private conversations.

Third, language reflects the range and depth of our experience, but our experience of the world is being impoverished to the extent that it is rendered artificial and prepackaged. Most of us no longer have the experience of skilled physical work on farms or in forests. Consequently words and metaphors based on intimate knowledge of soils, plants, trees, animals, landscapes, and rivers have declined. "Cut off from this source," Wendell Berry writes, "language becomes a paltry work of conscious purpose, at the service and the mercy of expedient aims" (1983, 33). Our experience of an increasingly uniform and ugly world is being engineered and shrink-wrapped by recreation and software industries and pedaled back to us as "fun" or "information." We've become a nation of television watchers and Internet browsers, and it shows in the way we talk and what we talk about. More and more we speak as if we are voyeurs furtively peeking in on life, not active participants, moral agents, or engaged citizens.

Fourth, we are no longer held together, as we once were, by the reading of a common literature or by listening to great stories and so cannot draw on a common set of metaphors and images as we once did. Allusions to the Bible and great works of literature no longer resonate because they are simply unfamiliar to a growing number of people. This is so in part because the consensus about what is worth reading has come undone. But the debate about a worthy canon is hardly the whole story. The ability to read serious literature with seriousness is diminished by overexposure to television and computers that overdevelop the visual sense. The desire to read is jeopardized by the same forces that would make us a violent, shallow, hedonistic, and materialistic people. As a nation we risk coming undone because our language is coming undone and our language is coming undone because one by one we are being undone.

The problem of language is a global problem. Of the roughly 5,000 languages now spoken on earth, only 150 or so are expected to survive to the year 2100. Language everywhere is being whittled down to the dimensions of the global economy and homogenized to accord with the imperatives of the information age. This represents a huge loss of cultural information and a blurring of our capacity to understand the world and our place in it. And it represents a losing bet that a few people armed with the words, metaphors, and mindset characteristic of industry and technology that flourished destructively for a few decades can, in fact, manage the earth, a different, more complex, and longer-lived thing altogether.

Because we cannot think clearly about what we cannot say clearly, the first casualty of linguistic incoherence is our ability to think well about many things. This is a reciprocal process. Language, George Orwell once wrote, "becomes ugly and inaccurate because our thoughts are foolish, but the slovenliness of our language makes it easier for us to have foolish thoughts" (1981, 157). In our time the words and metaphors of the consumer economy are often a product of foolish thoughts as well as evidence of bad language. Under the onslaught of commercialization and technology, we are losing the sense of wholeness and time that is essential to a decent civilization. We are losing, in short, the capacity to articulate what is most important to us. And the new class of corporate chiefs, global managers, genetic engineers, and money speculators has no words with which to describe the fullness and beauty of life or to announce its role in the larger moral ecology. They have no metaphors by which they can say how we fit together in the community of life and so little idea beyond that of self-interest about why we ought to protect it. They have, in short, no language that will help humankind navigate through the most dangerous epoch in its history. On the contrary, they will do all in their power to reduce language to the level of utility, function, management, self-interest, and the short term. Evil begins not only with words used with malice; it can begin with words that merely diminish people, land, and life to some fragment that is less than whole and less than holy. The prospects for evil, I believe, will grow as those for language decline.

We have an affinity for language, and that capacity makes us human. When language is devalued, misused, or corrupted, so too are those who speak it and those who hear it. On the other hand, we are never better than when we use words clearly, eloquently, and civilly. Language does not merely reflect the relative clarity of mind; it can elevate thought and ennoble our behavior. Abraham Lincoln's words at Gettysburg in 1863, for example, gave meaning to the terrible sacrifices of the Civil War. Similarly, Winston Churchill's words moved an entire nation to do its duty in the dark hours of 1940. If we intend to protect and enhance our humanity, we must first decide to protect and enhance language and fight everything that undermines and cheapens it.

What does this mean in practical terms? How do we design language facility back into the culture? My first suggestion is to restore the habit of talking directly to each other—whatever the loss in economic efficiency. To that end I propose that we begin by smashing every device used to communicate in place of a real person, beginning with answering machines. Messages like "Your call is important to us" or "For more options, please press five, or if you would like to talk to a real person, please stay on the line" are the death rattle of a coherent culture. Hell, yes, I want to talk to a real person, and preferably one who is competent and courteous!

My second suggestion is to restore the habit of public reading. One of my very distinctive childhood memories was attending a public reading of Shakespeare by the British actor Charles Laughton. With no prop other than a book, he read with energy and passion for two hours and kept a large audience enthralled, including one eight-year-old boy. No movie was ever as memorable to me. Further, I propose that adults should turn off the television, disconnect the cable, undo the computer, and once again read good books aloud to their children. I know of no better or more pleasurable way to stimulate thinking, encourage a love of language, and facilitate the child's ability to form images.

Third, those who corrupt language ought to be held accountable for what they do—beginning with the advertising industry. In 1997 the advertising industry spent an estimated $187 billion to sell us an unconscionable amount of stuff, much of it useless, environmentally destructive, and deleterious to our health. They fuel the fires of consumerism that are consuming the earth and our children's future. They regard the public with utter contempt—as little more than a herd of sheep to be manipulated to buy anything at the highest possible cost and at any consequence. Dante would have consigned them to the lowest level of hell, only because there was no worse place to put them. We should too. Barring that excellent idea, we should insist that they abide by community standards of truthfulness in selling what they peddle, including full disclosure of what the products do to the environment and to those who use them.

Fourth, language, I believe, grows from the outside in, from the periphery to center. It is renewed in the vernacular where human intentions intersect particular places, circumstances, and by the everyday acts of authentic living and speaking. It is, by the same logic, corrupted by contrivance, pretense, and fakery. The center where power and wealth work by contrivance, pretense, and fakery does not create language so much as exploit it. To facilitate control, it would make our language as uniform and dull as the interstate highway system. Given its way, we would have only one newspaper, a super-*USA Today*. Our thoughts and words would mirror those popular in Washington, New York, Boston, or Los Angeles. From the perspective of the center, the merger of ABC and Disney is okay because it can see no difference between entertainment and news. To preserve the vernacular places where language grows, we need to protect the independence of local newspapers and local radio stations. We need to protect local culture in all of its forms from domination by national media, markets, and power. Understanding that cultural diversity and biological diversity are different faces of the same coin, we must protect those parts of our culture where memory, tradition, and devotion to place still exist.

Finally, because language is the only currency wherever men and women pursue truth, there should be no higher priority for schools, colleges, and universities than to defend the integrity and clarity of language in every way possible. We must instill in our students an appreciation for language, literature, and words well crafted and used to good ends. As teachers we should insist on good writing. We should assign books and readings that are well written. We should restore rhetoric, the ability to speak clearly and well, to the liberal arts curriculum. Our own speaking and writing ought to demonstrate clarity and truthfulness. And we, too, should be held accountable for what we say.

In terms of sheer volume of words, factoids, and data of all kinds, this is surely an information age. But in terms of understanding, wisdom, spiritual clarity, and civility, we have entered a darker age. We are drowning in a sea of words with nary a drop to drink. We are in the process of committing what C. S. Lewis once called "verbicide" (Aeschliman 1983, 5). The volume of words in our time is inversely related to our capacity to use them well and to think clearly about what they mean. It is no wonder that during a dreary century of gulags, genocide, global wars, and horrible weapons, our use of language was dominated by propaganda and advertising and controlled by language technicians. "We have a sense of evil," Susan Sontag has said, but we no longer have "the religious or philosophical language to talk intelligently about evil" (Miller 1998, 55). That being so for the twentieth century, what will be said at the end of the twenty-first century, when the stark realities of climatic change and biotic impoverishment will become fully apparent? Can we summon the clarity of mind to speak the words necessary to cause us to do what in hindsight will merely appear to have been obvious all along?

6

Technological Fundamentalism

The implied objective of "progress" is—not *exactly* perhaps,
the brain in the bottle, but at any rate some frightful subhuman
depth of softness and helplessness.

—*George Orwell*

Scene 1: Entry to a classroom building. With a deafening noise he
revved up the two-cycle engine on a blower preparing to clean the
leaves, paper, and cigarette butts that had accumulated in the entry-
way. He made considerable progress herding the debris away from the
building and down the sidewalk until cigarette butts lodged in the
seams in the concrete. Turning, he blasted the miscreant trash at right
angles, but this only blew the debris onto the grass, posing still greater
difficulties. Moving cigarette butts and bits of paper in an orderly
fashion through grass is a challenge, even for a machine capable of
generating gale-force winds. Then the apparatus stalled out—"down
time," it's called. In that moment of sweet silence, I walked over and

inquired whether he thought a broom or rake might do as well. "What'd you say?" he responded. "Can't hear anything, my ears are still ringing!" I repeated the question. "S'pose so," he said, "but they think I'm more productive with this piece of *&!@."

Perhaps he is more productive. I do not know how experts calculate efficiency in complex cases like this. If, however, the goal is to disrupt public serenity, burn scarce fossil fuels, create a large amount of blue smoke, damage lung tissue, purchase expensive and failure-prone equipment, frazzle nerves, interrupt conversations, and improve the market for hearing aids, rakes and brooms cannot compete. When the technology and the task at hand are poorly matched, however, there is no real efficiency. In such cases the result, in Amory Lovins's telling phrase, is rather like "cutting butter with a chain saw."

Scene 2: Committee meeting. I once served on what is called with some extravagance the Educational Plans and Policies Committee. It is a committee to which one is elected, or sentenced, depending on your view. In one meeting we were casually asked to pronounce our blessing on a plan to link the entire campus so that everyone would be able to communicate with everyone else via computer, 24 hours a day, without leaving dormitory rooms or offices. This, we were told, was what our competitor colleges were doing. We were assured that this was the future. Information, we were informed, is doubling every six months. Electronic networking was judged to be an adequate response to that condition of information overload. Curious, I inquired what was known about the effects of computers on what we and our students think about or how well we can think about it. In other words, are there some things worth thinking about for which computers are ill suited? Can computers teach us to be properly skeptical of computers? Would people so wired and networked still want to talk to each other face to face? Would they remember how? Would they be sane? Or civil? Would they still know a tree from a bird? And after all the hype, what is the relation between information, knowledge, and wisdom? My fellow committee members, thoughtful persons all, stirred impatiently. After an awkward pause, one said, "We've been through this before and don't need to rehash the subject." I asked, "When?" Another awkward pause. No one could recall when that momentous conversation had occurred. "Well, it's all in the literature," said another. I asked for citations. None were forthcoming. What I had read on the subject by Joseph Weizenbaum

(1976), Theodore Roszak (1986), Neil Postman (1992), and C. A. Bowers (1993, 2000) would suggest to the curriculum committees of the world good reasons for caution. But these books had not been discussed by the committee, and no others were suggested.

Scene 3: Washington, D.C. A high public official is describing plans for the creation of a national information superhighway. The speech is full of high-tech words and "mega" this and that. Sober-looking public officials, corporate executives, and technicians glance at each other and nod approvingly. Members of the press dutifully scribble notes. TV cameras record the event. The questions that follow are mostly of the "gee whiz" kind. From the answers given, one might infer that the rationale for a superhighway is: (1) it will make the American economy more "competitive" because lack of information is what ails us; and (2) it's inevitable and can't be stopped anyway.

I am neither for nor against leaf blowers, computers, networks, or the information age, for that matter. My target is fundamentalism, which is not something that happens just to religious zealots. It can happen to well-educated people who fail to ask hard questions about why we do what we do, how we do it, or how these things affect our long-term prospects. We, leaf blowers and computer jockeys alike, have tended to become technological fundamentalists, unwilling, perhaps unable, to question our basic assumptions about how our tools relate to our larger purposes and prospects.

Scene 1 is an obvious case of technological overkill in which means and ends are not well matched. The deeper problem, noted by all critics of technology, from Mary Shelley and Herman Melville on, is that industrial societies are long on means but short on ends. Unable to separate can do from should do, we suffer a kind of technological immune deficiency syndrome that renders us vulnerable to whatever can be done and too weak to question what it is that we should do.

In scene 2, the committee did not know how computers affect what we pay attention to and how this, in turn, affects our long-term ecological prospects. Not knowing these things and being unwilling to admit them as honest, even important, questions, we did not know whether all of this technology could be used for good or not. Assuming that it could be used to good effect, we did not know how to do so. Seduced by convenience, dazzled by cleverness, armed with no adequate philosophy of technology, and not wanting to appear to our peers as premodern, we were at the mercy of those selling "progress" to us without a whisper about where it will ultimately take us.

In scene 3, much of the same is true on a larger scale as we approach the entry ramp of the information superhighway. Smart and well-meaning people believe this to be the cat's meow. But by what standard should we judge this enterprise? Will it, on balance, help us preserve biotic potential? Will it help to make us a more sane, civil, and sustainable culture? In this regard it is enlightening to know that a substantial part of the traffic now appearing on the superhighway so far built has to do with the distribution of pornography. Furthermore, the phrase "information superhighway" invites comparison to the interstate highways built in the United States between 1956 and the present. Any fair accounting of the real costs of that national commitment would include the contributions of the interstate system to the following problems:

- damage to urban neighborhoods and communities
- highway deaths
- loss of biological diversity
- damage to fragile landscapes
- urban sprawl
- polluted air
- acid rain
- noise pollution
- global warming
- destruction of an extensive national railway system
- distortion of American political life by an automobile lobby
- the foreign policy consequences of dependence on imported oil.

We, the children of the people who made or acquiesced in that decision, might prefer that these costs had been forthrightly discussed in 1956. Years from now, what might our children and grandchildren wish we had thought about before we built an information superhighway? We cannot know for certain, but we might guess that they would want us to have asked some of the following questions.

First, they might wish that we had been clearer about the purposes of the information superhighway. What problem was it intended to solve? What was the master idea behind it, and how might it support or undermine other master ideas in Western culture having to do with justice, fairness, tolerance, religious freedom, and democracy (Roszak 1986, 91–95)? Looking back, the rationale behind the interstate highway system was never much debated. To the contrary, it was presented as a combination of "national security" and "economic competitiveness," phrases that for nearly 50 years have been used to foreclose debate and conceal motives that should have been publicly examined.

Second, our descendants may wonder why we were so mesmerized by the capacity to move massive amounts of information at the speed of light. What kind of information for what purposes needs to be moved in such great quantities at that speed? At what velocity and volume does information become knowledge? Or wisdom? Is it possible that sometimes wisdom works inversely to velocity and volume? The bottleneck in this system will always be the space between our two ears. At what rate can we process information, or sift through the daily tidal wave of information to find that which is important or even correct? It would seem sensible to move the smallest possible amount of information consonant with the largest possible ends at a speed no faster than the mind can assimilate it and use it to good purpose. This speed is probably less than that of light. As discussed in chapter 3, the most valuable information relative to our long-term ecological prospects may prove to be that which is accumulated slowly and patiently—the kind of information that is mulled over and sometimes agonized over and with the passage of time may become cultural wisdom.

Third, future generations may wish that we had asked about the distribution of costs and benefits from the information superhighway. Looking back, the interstate highway system was a great boon to the heavy construction industry, car makers, oil companies, insurance companies, and tire makers. It was less useful to those unable to afford cars, who once relied on trains or buses. It was decidedly not beneficial to those whose communities were bulldozed or bisected to make way for multiple-lane expressways. Nor was it useful to those who had to spend a significant part of their lives driving to their newly dispersed workplaces. Accordingly, our descendants might wish us to ask whether access to the information superhighway will be fair? Will it be equally open to the poor? Will it be used to make society more or less equitable? Or more sustainable? Or will it be said of the information superhighway that it, like the "computer, as presently used by the technological elite, is . . . an instrument pressed into the service of rationalizing, supporting, and sustaining the most conservative, indeed reactionary, ideological components of the current Zeitgeist" (Weizenbaum 1976, 250)?

Our descendants will also wish that we had asked who will pay for the information superhighway. By one estimate, automobiles receive about $300 billion in various public subsidies each year (Nadis & MacKenzie 1993). They are supported by public road-building revenues, various taxes and tax loopholes, and by Defense Department expenditures to prepare for and fight wars to guarantee our access to oil. Might the same be true of the costs of the information superhighway?

Fourth, our descendants may wish that we had asked whether the standardization and uniformity imposed by information technology will homogenize our thoughts and language as well. For comparison, automobiles, interstate highways, and their consequences have served to homogenize American culture. Because of the scale of our automobility, our economy is less diverse and less resilient than it otherwise might have been. Our landscape has been rendered more uniform and standard to accommodate 200 million cars and trucks. Highways and automobiles have exacted a sizable toll on wildlife and biological diversity. Automobiles destroyed other and slower means of mobility including walking and bicycling. Will the imperatives of the information superhighway have analogous effects on our mindscapes? Will standardization and uniformity, shaped to fit information technology, homogenize our thoughts and language as well? Can cultural differences or cultural diversity survive technological homogenization? Will the vernacular information of indigenous cultures survive the information superhighway? Can increasingly uniform and standardized societies protect cultural diversity? And if they cannot, can they protect biological diversity?

The twentieth century is littered with failed technologies, once believed to be good in their time and promoted by smart and well-meaning people. The purveyors of automobiles, H-bombs, chlorinated fluorocarbons, toxic chemicals, and television all promised great things. These failed in large part because they succeeded too well. They became too numerous, or too efficient at doing one thing, or intruded too fully in places where they were inappropriate. A world with 100 million automobiles, for example, is probably okay. One with 500 million cars has more problems than I can list and fewer options for solving them than one might wish. Moreover, each of these technologies has caused unforeseen ecological and social problems that we wrongly call "side effects." There are, however, no such things as side effects, for the same reason that many technological accidents, as sociologist Charles Perrow (1984) once pointed out, are "normal accidents." Given human errors and acts of God, all such happenings are predictable events. What some call side effects of technology are the fine print of the deal when we think we are buying only convenience, speed, security, and affluence.

For a technological society, Garrett Hardin's (1968) query "what then?" is the ultimate heresy. But, standing, as we do, before such technological choices as nanotechnologies, genetic engineering, virtual reality machines, and information superhighways, no previous society needed its heretics more than ours. Information superhighways: What then? Ultimately, minds and perceptions so modified have different ecological prospects. Stripped of all the hype, the information superhighway is only a more complex, extensive, and expensive way to converse. But conversations conducted on that highway must ultimately be judged, as all conversations must be judged, not on the amount of talk or its speed, but by their intelligence, wisdom, and by what they inspire us to do.

7

Ideasclerosis

Let us first worry about whether man is becoming more stupid, more credulous, more weak-minded, whether there is a crisis in comprehension or imagination.

—*Paul Valery*

The time between innovations in technology and new products introduced into markets has steadily declined so that what had once taken decades has been reduced to months or a few weeks. As a result, we now have less time than ever to consider the effects of various innovations or systems of technologies on any number of other things, including our longer-term prospects. Contrast this pace, driven by the frenetic search for profit or power, with the rate of innovation in those things that would accrue to our long-term ecological health. This difference captures an important dimension of the problem of human survival in the twenty-first century. While we introduce new computing equipment every few months, we still farm in ignorance of

Charles Darwin and Albert Howard. Land-use thinking has barely begun to reckon with the thought of Aldo Leopold. After hundreds of studies on the potential for energy efficiency, our use of fossil energy, if somewhat more efficient, continues unabated. In short, innovations that produce fast wealth, whatever their ecological or human effects or impact on long-term prosperity, move ever more quickly from inception to market, while those having to do with human survival move at a glacial pace if they move at all. Why?

One possibility is that we are buried in an avalanche of information and can no longer separate the critically important from that which is trivial or perhaps even dangerous. This is certainly true, but it still does not explain why some kinds of ideas move quickly while others are ignored. Exhausted by consumption and saturated by entertainment, perhaps we have become merely "a nation of nitwits" (Herbert 1995) no longer willing or able to do the hard work of thinking about serious things. "The American citizen," Daniel Boorstin once wrote, "lives in a world where fantasy is more real than reality" (Boorstin [1961] 1978, 37). A casual survey of talk radio, television programs, and World Wrestling Federation events would lead one to believe this to be true as well. But, again, it does not explain why ecologically important ideas fail to excite us as much as contrived ones. Maybe the problem lies in the political arena, now dominated by wealthy corporations. Only those ideas that reinforce the power and wealth of the already powerful and rich succeed; all others are consigned to oblivion. This, too, is transparently obvious, but fails to explain why we are so easily entrapped by those with bad ideas. Maybe the problem is simply public cynicism, of which there is much evidence. Or perhaps we have simply created a very clever but ecologically stupid civilization. Indeed, as Kenneth Boulding once noted, it is difficult to overestimate stupidity in human affairs and its acceleration in recent decades. But that, too, merely begs the question.

Possibly the flow of ecologically sound ideas is blocked by the social equivalent of a logjam in a river. Again, there is plausible evidence for this possibility. Beginning in the late nineteenth century, for example, the industrial age spawned gargantuan organizations with simple goals, roughly analogous to the body/brain ratio of the dinosaur. Industrial behemoths such General Motors, similarly, lacked the wherewithal to think much beyond business equivalents of ingestion and procreation. Consequently, the ideas that flourished in organizations with great mass and single focus were the sort that increased either the scale or velocity of one thing or another in order to better serve the purposes of pecuniary accumulation, convenience, and power. The monomania of big organizations drove out thought for the morrow, warped lives, disfigured much of the world, and dominated the intellectual landscape. As a result, some of us live more conveniently, but the world is more toxic, dangerous, and far less lovely than it might otherwise be. Nonetheless, that model shaped our thinking about the proper organization of human affairs. Industrial-era organizations and industrialized societies lacked reliable means of appraising the collateral effects of their actions, what is called "feedback." And as Donella Meadows has noted, systems lacking feedback are by definition dumb. At a large enough scale, they are also dangerous.

But in societies dominated by large organizations, some kinds of ideas still spread like wildfire. Later generations will be hard pressed to explain the ferocious spread of nazism, communism, and various kinds of militant fundamentalism in the twentieth century (Conquest 1999). For such deranged ideas humans slaughtered each other by the millions. Our descendants, if not intellectually and morally impaired, will study the virulence of our ideologies much as we now study the etiology of disease. They will be astonished by our devotion to any number of other bad ideas such as the doctrine of mutual assured destruction. Most likely they will come to view our violence and political cupidity as a form of criminal insanity.

In one way or another, the dominant ideas of the twentieth century fit a pattern that political scientist James C. Scott calls "high-modernist ideology," which is "best conceived as a strong, one might even say muscle-bound, version of the self-confidence about scientific and technical progress, the expansion of production, the growing satisfaction of human needs, the mastery of nature (including human nature), and above all, the rational design of social order commensurate with the scientific understanding of natural laws" (1998: 4). Taken to its extreme, devotees of high modernism, in Scott's words, "were guilty of hubris, of forgetting that they were mortals" (ibid., 342). Whether in forestry, agriculture, urban planning, or economics, the practice of high modernism meant excluding qualitative and subtle aspects of rural places, natural systems, cities, and people in order to maximize efficiency, control, and economic expansion. The acolytes of the faith steadfastly hold to a vision of humankind become godlike, transcending all limitations including death. When it is all said and done I doubt that, on balance, high modernism will have eliminated much suffering. But it will have served to anesthetize our higher sensibilities and drastically deflect human nature or eliminate humans altogether. Indeed, the latter is the stated goal of all of those intrepid pioneers in the brave new sciences of virtual reality and artificial intelligence, who regard the displacement of humans by superior and self-replicating devices as an evolutionary mandate.

Given the present momentum of research, twenty-first-century technologies, notably genetics, nanotechnologies, and robotics, will change what it means to be human. They may well threaten human survival. In the words of software engineer Bill Joy (2000, 242), "we are on the cusp of the further perfection of extreme evil." We are driven "by our habits, our desires, our economic system, and our competitive need to know," but we have "no plan, no control, no brakes" (ibid., 256). Joy believes that the "last chance to assert control . . . is rapidly approaching" (ibid.). Others such as Ray Kurzweil, author of *The Age of Spiritual Machines*, counsel resignation because these changes are "inexorable" and "inevitable" (1999, 253).

Looking ahead, as best we are able, what can be said about the trajectory of human intelligence? Is it possible to harness intelligence to purposes that demean it? Is it possible to create conditions that are hostile to sober reflection, decency, and foresight? We have good reasons to think that the conditions that nurture ecologically solvent ideas and wisdom are mutable, fragile, and increasingly threatened by the march of mere cleverness and the avalanche of artifice and sensation on the human psyche. And we now know that it may well be possible to destroy human intelligence altogether by creating a form of superior intelligence that could well regard us as a nuisance to be removed.

It is against the intoxication of high modernism which conservation biologists and their allies struggle. In the blizzard of technological possibilities, how do we cultivate what Aldo Leopold once called a "refined taste in natural objects" or a "striving for harmony with land" (1953, 150, 155)? How do we create the intellectual and moral capital for a "society decently respectful of its own and all other life, capable of inhabiting the Earth without defiling it" (Leopold 1999,

318–319)? What ecologically grounded alternative to high modernism do we offer? How do we quickly capture the imagination of the general public for the slow things that accrue to the health of the entire land mechanism?

It is far easier to describe the general content of such ideas than how they might become powerful in a consumer culture. In one way or another, the ideas we need would extend our sense of time to the far horizon, broaden our sense of kinship to include all life forms, and encourage an ethic of restraint. Not one of these can be hurried into existence. This is not first and foremost a research challenge as much as it is a kind of growing up. It is perhaps more like a remembering of what Erwin Chargaff (1980, 47) once called "old and solid knowledge" that has existed in those times and places where foresight and compassion were cultivated. A culture permeated with old and solid knowledge makes no fetish of novelty and so does not suffer the cultural equivalent of amnesia. The perennial wisdom of humanity honors mystery and acknowledges the need for caution and large margins. It knows that human intelligence is always and everywhere woefully inadequate and that we need large margins. Much of this old and ecologically sound knowledge is embedded in scriptures, law, literature, and ancient customs. But how is this to be made vivid for an entire culture suffering from attention deficit disorder?

Broadly speaking, I think we have three general strategies. One is to try to capture public imagination by dramatizing aspects of our situation. The Clock of The Long Now Foundation, for example, intends to create a 10,000-year clock that "ticks once a year, bongs once a century, and [from which] the cuckoo comes out every millennium" (Brand 1999, 3). To counter the hypernervousness of the nanosecond culture, Stewart Brand and his colleagues intend to create something comparable to the photograph of Earth from the Apollo spacecraft. The goal is to revolutionize our sense of time from the short term (*kairos*) to the long term (*chronos*), from cleverness to wisdom (ibid., 9). The actual experience of this device, whatever it might be, they describe as "Whew, *Time!* And me in it . . . like coming upon the Grand Canyon by surprise" (ibid., 49). Perhaps focusing on the longer sweep of time would make more of us amenable to precautionary steps to preserve those things essential to the long now and less susceptible to the political, technological, and economic contagions of the moment. On the other hand, people accustomed to being entertained might regard it only as another theme park—a sort of Disneyland. And some things, such as soil and biological diversity, cannot be dramatized so easily.

A second strategy is aimed at changing how we see the world by creating more accurate and telling metaphors and theories. *Natural Capitalism* by Paul Hawkin, Amory Lovins, and Hunter Lovins (1999), for example, is a painstaking and compelling case for including ecological capital in our economic accounting and business practices. They propose to reconcile the economy to fit the realities of natural systems by pointing out the logical inconsistencies in our current modes of thinking. Indeed, a great deal of environmentalism is an attempt to change mental models and perspectives to break the chains of anthropomorphism. But changing minds and paradigms is a slow business, proceeding, when it does, mostly funeral by funeral as one generation gives way to the next. The powers of denial are everywhere strong and deeply entrenched, but given time metaphors can change and ideas do spread.

The third strategy, political change, has fallen into disrepute in the age of hypercapitalism. In our pursuit of fast wealth, we allowed ourselves to be bamboozled into believing that government was the problem. As a result, the public sector, relative to multinational corporations, has been weakened virtually everywhere. While capitalism is triumphant, there is a deficit of political ideas and an atrophy of the sense of common interests and community. At the very time we need robust political ideas to confront unprecedented changes in technology, increasing concentration of wealth, rising human needs, and serious environmental threats, we find political confusion, vacillation, and mendacity. The kind of political leadership we need has yet to appear. But the ideas necessary for a solvent future are relatively straightforward. We must create the same kind of separation between money and politics that we once established between church and state. And we must create the political capacity to protect the integrity of earth systems and biodiversity and thereby the legitimate interests of our descendants. This requires, in turn, the capacity to exert farsighted public control over capital and economic power. It is no easy thing to do, but doing it is far easier than not doing it.

The success of these strategies, in turn, hinges on whether the public is educated and equipped to comprehend such things. But at the time when we need a larger idea of education, our proudest research universities, almost without exception, have aspired to become the research and development wing of high modernism. The Bayh-Dole Act of 1980 permitted universities to patent results of federally funded research (Press and Washburn 2000, 41). Combined with the decline of defense spending, the results have been dramatic. The more prestigious institutions have become partners, and sometimes accomplices, of major corporations in return for large contributions and contracts. Many have established offices to foster and administer the commercialization of research. Corporations increasingly dictate the terms of research and its subsequent use, thereby compromising the free flow of ideas and contaminating truth at the source. Unsurprisingly, research is mostly directed to areas that hold great financial promise, not to great human needs. There is seldom much financial profit in ideas pertaining to preservation of biological diversity, land health, sustainable resource management, and real human improvement—precisely what we need most. And there is virtually never quick profit in turning out merely well-educated, thoughtful, and ecologically competent citizens.

It should be a matter of some embarrassment that the best ideas about the challenge of sustainability and appropriate responses to it have come disproportionately from people and organizations at the periphery of power and influence not from those at the center. Small nonprofit organizations are often the best source of ideas we have about the preservation of species, soil, people, places, local culture, and margins for error. It is time for institutions of higher education to catch up. It is time to reinvent higher education by breaking down all of those institutional and disciplinary impediments to the flow of ideas on which we might build a durable and decent civilization.

8

Ideasclerosis, Continued

If and when the ecological idea takes root, it is likely to change things.

—Aldo Leopold

General George Lee Butler ascended through the ranks of the air force from fighter pilot to the commander of the U.S. Strategic Command. He was a true believer in the mission of the military and specifically in the efficacy of nuclear deterrence, but he was also a thinking man, and his doubts had begun in the 1970s. Finally, in 1988 during a visit to Moscow, he wrote, "it all came crashing home to me that I really had been dealing with a caricature all those years" (Smith 1997, 20). Butler was nearing the end of what he described as a "long and arduous intellectual journey from staunch advocate of nuclear deterrence to a public proponent of nuclear abolition" (Butler 1996). The difference between Butler and many others in the military was that "he reflected on what he was doing time and again," and much of what he'd come to take for normal did not add up. He wrote, "We have yet to fully grasp the monstrous effects of these weapons . . . and the horrific prospect of a world seething with enmities, armed to the teeth with nuclear weapons." To do so will require overcoming a "terror-induced anesthesia which suspend[s] rational thought" in order to see that "we cannot at once keep sacred the miracle of existence and hold sacrosanct the capacity to destroy it" (Butler 1998). Butler, now in private business, devotes a substantial part of his life to the abolition of nuclear weapons.

Ray Anderson, founder and CEO of Interface Corporation, experienced an even more abrupt conversion. In 1994, after 21 years as the head of a highly successful carpet and tile company, he was asked by his senior staff to define the company's environmental policy. "Frankly," he writes, "I did not have a vision" (Anderson 1998, 39). In trying to develop one, he happened to read Paul Hawken's (1993) *The Ecology of Commerce*, and the effect was, as he put it, like "a spear in the chest" (Anderson 1998, 23). He subsequently read other books ranging from Daniel Quinn's *Ishmael* to Rachel Carson's *Silent Spring*. The effect of his reading and reflection was to deepen and intensify an emotional and intellectual commitment to transform the company. Anderson went on to define environmental goals for Interface that has placed the company in the forefront of U.S. business, a transformation that he describes as "a phenomenon of the first order" (Anderson 1998, 183). Instead of merely complying with the law, Anderson aims to make Interface a highly profitable, solar-powered company discharging no waste and converting used product into new product through what the company calls an "evergreen lease." The Interface annual report reads like a primer in industrial ecology written by thinkers like Paul Hawken, William McDonough, and Amory Lovins. Anderson, now in his midsixties, has become a tireless and eloquent advocate for the ecological transformation of business.

Butler and Anderson are extraordinary people. They were both at the top of their respective professions when they came to the realization that something fundamental was wrong. They were thoughtful and honest enough to eventually see through the complacency and pretensions that accumulate around organizations and institutions like barnacles on the hulls of ships. They are deeply religious men who saw the necessity for change in moral terms and had enough moral energy to transcend the world of cold calculation to see their professions in a larger human and humane perspective and enough courage to risk failure, rejection, and ridicule.

People like Butler and Anderson are threatening to the stability and smooth functioning of organizations and institutions. Butler's challenge to the defense establishment, an entity not famous for its encouragement of new ways of seeing things, is the more daunting. As the CEO of Interface, Anderson has considerably more leverage over outcomes. But both men represent the kind of professional that Donald Schon (1983) once called "the reflective practitioner." In Schon's words, the reflective practitioner is inclined to engage "messy but crucially important problems" through a process that combines "experience, trial and error, intuition, and muddling through" (ibid., 43). Moreover, the reflective practitioner

> allows himself to experience surprise, puzzlement, or confusion in a situation which he finds uncertain or unique. He reflects on the phenomena before him, and on the prior understandings which have been implicit in his behavior. He carries out an experiment which serves to generate both a new understanding of the phenomena and a change in the situation. [He] is not dependent on the categories of established theory and technique . . . his inquiry is not limited to a deliberation about means which depends on a prior agreement about ends. He does not keep means and ends separate . . . he does not separate thinking from doing. (Schon 1983, 68)

In contrast, most professionals are "locked into a view of themselves as technical experts, find nothing in the world of practice to occasion reflection [having] become too skillful at techniques of selective inattention, junk categories, and situational control" (ibid., 69). For them, professionalism functions, as Abraham Maslow once described science, "as a Chinese Wall against innovation, creativeness, revolution, even against new truth itself if it is too upsetting" (1966, 33). But organizations and institutions do not often reward mavericks who upset rules and procedures or who question the unquestionable. To the contrary, they are penalized, ostracized, or, worse, elaborately ignored because they threaten what are perceived to be core values and comfortable routines.

The problem that reflective practitioners face is that they mostly work in rigid organizations or professions that function unreflectively. Both Butler and Anderson challenged the fundamental worldview of their respective organizations by seeing the organization and its larger environment at a higher level of generality. From that vantage point Butler could see that nuclear weapons only compounded the problem of security, and Anderson could see the environmental and human havoc caused by a prosperous company otherwise doing everything by the rules. To accommodate people like Butler and Anderson, an organization must meet "extraordinary conditions" that include plac[ing] a high priority on flexible procedures, differentiated responses, qualitative appreciation of complex processes, and decentralized responsibility for judgment and action . . . mak[ing] a place for attention to conflicting values and purposes" (Schon 1983, 338). In short, an organization must be capable of learning (Schon 1971).

The concept of a learning organization sounds like an oxymoron, but the human prospect depends every bit as much on the capacity of organizations to learn as it does on individual learning. Few scholars have thought more deeply about the possibility and dynamics of organizational learning than Massachusetts Institute of Technology professor Peter Senge. According to Senge, learning organizations are those in which "people continually expand their capacity to create the results they truly desire, where new and expansive patterns or thinking are nurtured, where collective aspiration is set free, and where people are continually learning how to learn together" (1990, 3). Learning organizations, Senge writes, "develop people who learn to see as systems thinkers see, who develop their own personal mastery, and who learn how to surface and restructure mental models collaboratively" (ibid., 367). They foster people capable of seeing the organization and institution at a higher level of generality and thereby capable of challenging basic premises. In short, learning organizations encourage creativity, innovation, out-of-the-box thinking, and the heretics who speak to fundamentals. On such people and on such organizations the human future depends.

"For twenty centuries and longer," in Aldo Leopold's words, "all civilized thought has rested upon one basic premise: that it is the destiny of man to exploit and enslave the earth" (1999, 303). And we've got-

ten good at it, multiplying and becoming fruitful beyond the wildest
dreams of our ancestors. Throughout history we learned mostly
driven by necessity: failure, war, famine, overcrowding. Now we have
to learn entirely new things, not because we failed in the narrow sense
of the word, but because we succeeded too well. In one way or another all of the challenges of the twenty-first century are linked to the
fact that we've procreated too rapidly and produced more waste than
the earth can process. We suffer from a new dynamic of excess success and must make a rapid transition to a more restrained and elegant
condition called sustainability. To do so, what must we learn? We
must learn that we are inescapably part of what Leopold called "the
soil-plant-animal-man food chain" (ibid., 198). We must master systems dynamics, learning ideas of feedback, stocks, flows, and delays
between cause and effect. And we must learn to see ourselves as
trustees of the larger community of life, which is to say that we must
embrace a higher and more inclusive level of ethics. We must, in other
words, see the human enterprise and all of our own little enterprises
at a higher level of generality in a much longer span of time and re-
strain ourselves accordingly. Who will teach us these things?

The fact is that much or even most of what we've learned about
this transition has been through the efforts of organizations not usually regarded as educational and by mavericks operating as reflective
practitioners against the grain of their professions. Some of the best
work on ecological technology, for example, occurs in places like
Ocean Arks, Massachusetts, or Gaviotas, Colombia. The creative edge
in urban planning and design has been happening on the streets of
Curitiba, Brazil, or in cities like Chattanooga, Tennessee, or in new
developments like Village Homes in Davis, California, Haymount,
Virginia, or Prairie Crossings, Wisconsin. The best forestry management is being practiced in the forests of the Menominee tribe in
north-central Wisconsin. The most advanced thinking about energy
use and automobiles comes from the Rocky Mountain Institute in
Colorado. Some of the best thinking about applied economics is taking place at small institutions like Rethinking Progress, Inc., or The
Center for a New American Dream. We are learning industrial ecology from companies such as Interface, Inc., and 3-M. The best analysis of our global plight comes from institutions like the WorldWatch
Institute and the World Resources Institute.

But where, in the most critical and fateful period of human history, does one find the prestigious and well-endowed institutions of higher education? The short answer is that most have yet to summon the wherewithal and energy to do very much. Relative to the transition to sustainability, institutions of higher education are under-achievers.[1] On balance, then, it is unclear whether higher education will be a positive or negative factor in the transition ahead. What we do know is that higher education can, in Jonathan Kozol's words, "prosper next to concentration camps . . . collective hysteria, savagery—or simply quiet abdication in the presence of ongoing misery outside the college walls" (1985, 169). It has certainly adapted comfortably with the corporate dominated extractive economy that lies at the heart of our environmental and social problems. Why?

The problem stems, I think, from a deep-seated complacency that bears resemblance to the history of the U.S. auto industry. Consider that slow-moving, dim-witted colossus, General Motors circa 1970, that failed to check its rearview mirror. Toyota and Honda were in the passing lane. Our product, too, is often overpriced and of uncertain quality. We have lost our sense of direction, becoming all things to all people. Long ago we surrendered the idea of guiding students to a larger vision of self and life in favor of merely well-paying careers. On the most important issues of the time, we have sounded an uncertain trumpet or no trumpet at all. We are being corrupted by financial dependence on corporate interests that have every intention of using higher education to their advantage. And a glance at the rearview mirror shows competitors such as the Internet, organizations offering distance learning, and other vendors coming up fast in the passing lane.

The question, then, is whether the institutions that purport to advance learning can themselves learn new ways appropriate for an ecological era. What would it mean for the ecological idea to take root in colleges and universities? It would mean, for one thing, that such institutions would have to become learning organizations in order to reinvent themselves. This requires rethinking institutional

---

1. Berea College, College of the Atlantic, Green Mountain College, Northland College, Prescott College, and Warren Wilson College are notable exceptions.

purposes and procedures at a higher level of generality. It would mean changing routines and old ways of doing things. It would require a willingness to accept the risks that accompany change. It would require a more honest accounting to include environmental costs. Instead of bureaucratic and academic fragmentation, the transition would require boundary crossing and systems ways of thinking and doing. Instead of being reactive organizations, they would become proactive, with an eye on the distant future. Instead of defining themselves narrowly, they would redefine themselves and what they do in the world at a higher and more inclusive level.

What do these things mean in everyday terms? For one thing, the transition to becoming a learning organization would change who has lunch with whom. The requirement for openness would tend to dissolve the barriers separating disciplines and encourage bolder, more imaginative, and more useful kinds of thought, research, and teaching. It would help to initiate a more honest dialogue about knowledge and its relation to our ecological prospects. The transition would require rethinking the standards for academic success to encourage engagement with real and sometimes messy public problems. It would expand the definition of our "product" from courses taught and articles published to include practical problem solving. It could change how we define our clientele in order to educate, and be educated by, a wider constituency. It would change the standards against which we evaluate institutions of higher education to include our real ecological impacts on the world and perhaps those of our graduates. Since learning, both institutional and individual, begins with an ability to see things in perspective, organizational learning might serve to deflate the pomposity that often pervades the upper echelons of the academy. Finally, transitions don't often occur without leadership, and higher education needs leaders as bold, honest, and capable as George Lee Butler and Ray Anderson.

It is not whether higher education will be reinvented, but rather who will do the reinventing and to what purposes. If we fail to make institutions of learning into learning organizations, others will reinvent the academy for less worthy purposes. If we fail to elevate professional standards, those professions will be irrelevant to the transition ahead, or worse, an impediment. If we, in higher education, cannot make these changes, the possibility that the great transition ahead will be informed by liberally educated people will also decline. That means, in short, that the ideas necessary for a humane, liberal, and ecologically solvent world will be lost in favor of a gross kind of global utilitarianism.

§ 3

THE POLITICS OF DESIGN

9

None So Blind: The Problem of Ecological Denial

None so blind as those that will not see.

—*Mathew Henry*

Willful blindness has reached epidemic proportions in our time. Nowhere is this more evident than in recent actions by the U.S. Congress to deny outright the massive and growing body of scientific data about the deterioration of the earth's vital signs, while attempting to dismantle environmental laws and regulations. But the problem of ecological denial is bigger than recent events in Congress. It is flourishing in the "wise use" movement and extremist groups in the United States, among executives of global corporations, media tycoons, and

David Ehrenfeld coauthored this chapter.

on main street. Denial is in the air. Those who believe that humans are, or ought to be, something better than ecological vandals need to understand how and why some people choose to shun reality.

Denial, however, must be distinguished from honest disagreement about matters of fact, logic, data, and evidence that is a normal part of the ongoing struggle to establish scientific truth. Denial is the willful dismissal or distortion of fact, logic, and data in the service of ideology and self-interest. The churchmen of the seventeenth century who refused to look through Galileo's telescope, for example, engaged in denial. In that instance, their blind obedience to worn-out dogma was expedient to protect ecclesiastical authority. And denial is apparent in every historical epoch as a willing blindness to the events, trends, and evidence that threaten one established interest or another.

In our time, great effort is being made to deny that there are any physical limits to our use of the earth or to the legitimacy of human wants. On the face of it, the case is absurd. Most physical laws define the limits of what it is possible to do. And all of the authentic moral teachings of 3,000 years have been consistent about the dangers and futility of unfettered desire. Rather than confront these things directly, however, denial is manifested indirectly.

A particularly powerful form of denial in U.S. culture begins with the insistence on the supremacy over all other considerations of human economic freedom manifest in the market economy. If one chooses to believe that economies so dominated by lavishly subsidized corporations are, in fact, free, then the next assumption is easier: the religious belief that the market will solve all problems. The power of competition and the ingenuity of technology to find substitutes for scarce materials, it is believed, will surmount physical limits. Markets are powerful institutions that, properly harnessed, can accomplish a great deal. But they cannot substitute for healthy communities, good government, and farsighted public policies. Nor can they displace the laws, both physical and moral, that bound human actions.

A second indirect manifestation of ecological denial occurs when unreasonable standards of proof are required to establish the existence of environmental threats. Is the loss of species a problem? Well, if you think so, just name one species that went extinct today! The strategy is clear: focus on nits, avoid large issues, and always demand an unattainable level of proof for the existence of any possible problem before agreeing to any action to forestall potential catastrophe.

True, no such standards of proof of likely Soviet aggression were required to commit the United States to a $300 billion defense budget. But denial always works by establishing double standards for proof.

Third, denial is manifest when unwarranted inferences are drawn from disconnected pieces of information. For example, prices of raw materials have declined over the past century. From this, some have drawn the conclusion that there can be no such thing as resource scarcity. But the prices of resources are the result of complex interactions between resource stocks/reserves, government subsidies, unpriced ecological and social costs of extraction, processing, transportation, the discount rate, and the level of industrial growth (which turned down in the 1980s). This is why prices alone do not give us accurate information about depletion, nor do they tell us that the planetary sinks, including the atmosphere and oceans, are filling up with wastes they cannot assimilate.

Moreover, the argument from prices and other economic indicators does not take into account the sudden discontinuities that often occur when limits are reached. A typical example from physics is stated in Hooke's Law: Stress is proportional to strain, *within the elastic limit*. The length of an elastic band is proportional to the stretching force exerted on it—until the band snaps. In biology, the population crashes that sometimes occur when carrying capacity is reached provide another example. There are many more.

Fourth, denial is manifest in ridicule and ad hominem attacks. People inclined to think that present trends are not entirely positive are labeled doomsayers, romantics, apocalyptics, Malthusians, dreadmongers, and wackos. In a book that dominated environmental discussion on Earth Day 1995, *Newsweek* writer Gregg Easterbrook, for example, says that such people (whom he calls "enviros") "pine for bad news." They suffer from a "primal urge to decree a crisis" (1995, 440) and "subconscious motives to be alone with nature" (ibid., 481). Pessimism, for them, is "stylish." They are ridiculous people with nonsensical views, who do not deserve a serious response; this relieves those doing the name calling and denying from having to think through complex and long-term issues.

Fifth, denial is manifest in confusion over time scales. Again, Easterbrook spends the first 157 pages of his 698-page opus explaining why in the long view things such as climatic change and soil erosion are minor events. Shifting continents, glaciation, and collision with asteroids have wreaked far greater havoc than human-caused degradation. "Nature," he says, "has for millions of centuries been generating worse problems than any created by people" (1995, xvii). I do not for a moment doubt the truth of this assertion. Nor do I doubt that from, say, Alpha Centauri, a nuclear war on Earth would scarcely make the midday farm report. Easterbrook enjoins us to place our ecological woes in the perspective of geologic time, and from a sufficient distance they do indeed look like a quibble. The earth is a fortress, he says, capable of withstanding all manner of insult and technological assault. But we don't live on Alpha Centauri, and events that may be trivial in a million years loom very large to us with our 75-year life spans, our few-hundred-year-old countries, and our 8,000-year-old agricultural civilization.

Denial is manifest, sixth, when large and messy questions about the partisan politics of environmental issues are ignored. In the fall of 1994, about the same time that Easterbrook would have been working over the galley pages for his book, agents of the Republican party were drafting the final version of *The Contract with America*, a major goal of which was to dismantle all of the environmental laws and regulations so painstakingly erected over the past 25 years. Ecological optimism was blindsided by political reality.

Why is denial happening? It is happening, first, because in the face of serious problems such as the increasing gap between the rich and everyone else, and the related problems caused by unrestrained corporate power, we look for scapegoats rather than confront problems directly. Historian Richard Hofstadter once called this the "paranoid style of politics." Practitioners of paranoid politics use conspiracy theories to explain why things are not as good as they ought to be. Since the collapse of the Soviet Union, reliably awful enemies are more difficult to find. Accordingly, environmentalists, bureaucrats, gays, and ethnic minorities have replaced communists as the enemies of choice.

Second, and perhaps most obvious, denial is a defense against anxiety. Many of the environmental changes that are now happening are deeply disturbing, but they constitute only a part of the assaults on our well-being that most of us face daily. It is natural to want to lighten our load of troubles by jettisoning a few. Environmental problems are rarely as personally pressing as sickness or loss of a job, so out they go. This kind of denial can provide some immediate relief of anxiety. However, it merely delays the confrontation with ecological reality until the time when environmental events, breaking through the screen of denial, force themselves upon us. When that occurs, our ecological troubles will be far more painful and far less tractable to deal with than they are now.

Ecological denial is happening, third, because it seems plausible to the ill-informed. Polls show that only 44 percent of Americans believed that human beings developed from earlier species, while only 63 percent were aware that human beings negatively affect biodiversity. This was the lowest response among the citizens of 20 countries surveyed. People so ignorant are mere fodder for those who would harness denial for their own purposes.

Fourth, it may be fair to say that ecological denial is happening in the public because environmental advocates often appear to be elitist and overly focused on an ideal of pristine nature, to the exclusion of real people. We have not bridged the gap between environmental quality and class as imaginatively and aggressively as we ought to have done. As a result, many people see conservation biologists and environmental activists as members of yet another special interest group, not working for the general good. It is clear that we will have to do a better job explaining to the public why the environment is not an expendable concern unrelated to real prosperity and community. How is this to be done?

I would like to recommend the following steps. First, members of the conservation community must not deny that we live in a society which desperately needs fixing and in which denial is seductively easy and cheap, at least for a time. We must acknowledge and seek to understand the connection between poverty, social injustice, and environmental degradation. We must acknowledge and seek to understand the connection between rootlessness and environmental irresponsibility. We must acknowledge and seek to understand the connection between the loss of functional human communities and the inexorable decline in the state of the earth.

Second, we should take our critics seriously enough to read what they have to way. I recommend a close reading of books such as *But Is It True?* by the late Aaron Wildavsky (1995) and Ronald Bailey's edited volume called *The True State of the Planet* (1995). We need to separate those things on which we may agree from those on which we cannot agree, the plausible from the implausible, and be utterly clear about the difference.

Third, we should take words more seriously than we have in the past. Without much of a fight, we have abandoned words such as "progress," "prosperity," and "patriotism" to those who have cheapened and distorted their meanings beyond recognition. We need to take back the linguistic and symbolic high ground from the deniers. At the same time, however, some of us need to be much more careful about using apocalyptic words such as "crisis." "Crisis," a word taken from the field of medicine, implies a specific time in an illness when the patient hovers between life and death. But few environmental problems conform closely to that model. We do not doubt for a second that we now face some genuine crises and that we will face others in the future. But for the most part, ecological deterioration will be a gradual wasting away of possibilities and potentials, more like the original medical meaning of the word "consumption."

Finally, we should all learn to recognize the signs of ecological denial, so that when we see it in operation we can expose it for what it is and force an honest discussion of the real issues that deserve our immediate and full concern.

10

Twine in the Baler

I recall a true story about an Ozark farmer who telephoned his neighbors one fine June day asking for help in getting in his hay. Arriving at the hayfield, people found the farmer baling his hay, but without twine in the baler. Unbound piles of hay, which would have to be entirely reraked and rebaled, lay all over the field. The farmer, with a bottle of whiskey in his lap, was feeling no pain, as they say, and did not seem to notice the problem, nor did the dozen or so men, similarly anesthetized, standing around the pickup trucks at the edge of the field. Believing the lack of twine to be a serious problem, one of the volunteers, a newcomer to such haying operations, suggested putting a roll of twine in the baler. To which an old-timer replied: "Naw, no need for that. Ol' Billy-Hugh [the farmer in question] is having too much fun to stop now."

This story says something important about intention. Those of us who arrived on the scene ready to work failed to understand that the purpose of the event had nothing to do with getting in hay. This was a party, haying the pretext. Once we understood that, all of us could get in the flow, so to speak.

A good many things, including politics, work similarly. One of the best books ever written about politics, *The Symbolic Uses of Politics* (Edelman 1962), develops the thesis that the purpose of political activity is often not to solve problems but only to appear as if doing so. The politics of sustainability, unfortunately, provide no obvious exception to this tendency to exalt symbolism over substance. And of symbols and words there is no end. The subject of sustainability has become a growth industry. Government- and business-sponsored councils, conferences, and public meetings on sustainability proliferate, most of which seem to be symbolic gestures to allay public anxieties, not to get down to root causes. What would it mean to put twine in our baler? I would like to offer three suggestions.

Getting serious about the problem of sustainability would mean, first, raising difficult and unpolitic questions about the domination of the economy by large corporations and their present immunity from effective public control. All of the talk about making economies sustainable tends to conceal the reality that few in positions of political or economic power have any intention of making corporate power accountable to the public, let alone reshaping the economy to fit ecological realities. Free trade, as it is now proposed, will only make things worse. Scarcely any countervailing power to predatory capital exists at the national level, and none exists at the global level. In such a world, economic competitiveness will be the excuse for any number of egregious decisions that will be made by people who cannot be held accountable for their actions.

Putting twine in the baler in this instance would mean, among other things, enforcing limits on the scale of economic enterprises and undoing that piece of juristic mischief by which the Supreme Court in 1886 (*Santa Clara County v. Southern Pacific Railroad*) bestowed on corporations the full protection of the Bill of Rights and the Fourteenth Amendment, giving them, in effect, the legal rights of persons (Grossman and Adams 1993). That decision, and others subsequently, have placed U.S. corporations beyond effective public control. The right to use their wealth as persons enables them to influence the votes of legislators and to evade the law and weaken its administration. Exercising their right of free speech, corporations fill the airwaves with incessant advertisements that condition and weaken the public mind. The exercise of their economic power creates dependencies that undermine public resolve. Their sheer pervasiveness erodes the basis for alternative, and more sustainable, ways to provision society. The practical effect is that corporations are seldom motivated to do what is in the long-term interest of humanity if it costs them much. And were they to do so, their stockholders could sue them for failing to maximize returns to capital. It is hardly possible to conceive of any long-lived society that provisions itself by agents so powerful yet so unaccountable and so focused on short-term profit maximization. Twine in the baler would mean putting teeth in the charters of corporations in order to make them accountable over the long term and dissolving corporations for failure to abide by their terms.

Getting serious about sustainability, second, would require a radical reconsideration of the present laissez-faire direction of technology. Many advocates of sustainable development place great faith in the power of technology to improve the efficiency with which energy and resources are used. Better technology may well succeed in doing so, but the same unfettered development of technology has a darker side about which little is said. For example, Marvin Minsky (1994), in a recent issue of *Scientific American,* asked whether "robots will inherit the earth." His answer was an enthusiastic yes. He and others are, accordingly, working hard to "deliver us from the limitations of biology," intending to replace human bodies with mechanical surrogates and our brains with devices having the capacity to "think a million times faster than we do" (Minsky 1994, 112; Moravec 1988). Other knowledgeable observers predict that artificial intelligences "will eventually excel us in intelligence and it will be impossible to pull the plug on them. . . . They will be impossible to keep at bay. . . . Human society will have to undergo drastic changes to survive in the face of artificial intelligences. . . . Their arrival will threaten the very existence of human life as we know it" (Crevier 1994, 341). True or not, many believe such things are possible, desirable, or merely inevitable, and that belief means that such things will almost certainly be attempted. But do we really want some research scientists—for the sake of profit, fame, or just the sheer fun of it—to create machines with the potential to displace the rest of us and our children? Who has given them the right to threaten the existence of human life?

Little or no public effort is being made to question whether we want to go where technologies such as artificial intelligence, nano-technologies, genetic engineering, or virtual reality are taking us. Nor do we have the institutions necessary to weigh the consequences of technological change against alternative paths of development. Modern society is approaching the future with the throttle of technological change jammed to the floor, and the issue of slowing and directing it is not on the public agenda in any coherent way. Putting twine in the baler in this instance would mean admitting that technological choices are often political choices that affect the entire society. As political decisions, such choices should be made in an open and democratic manner in participatory institutions capable of evaluating technological choices as thoroughly as possible against alternatives that may accomplish better results more cheaply and with fewer side effects.

Getting serious about the crisis of sustainability will mean, third, a considerable change in how we think about our responsibilities as citizens. On one side of the issue are those who believe that environmental policy must be based solely on rational self-interest, not on appeals to moral behavior. "Whenever environmentalism has succeeded," they argue, "it has done so by changing individual incentives, not by exhortation, moral reprimand, or appeals to our better natures" (Ridley and Low 1993, 80). Certainly, public policies ought to tap self-interest whenever possible, but proponents often go beyond this truism to say something more sweeping about human potentials and, by implication, the nature of the emergency ahead. At the core of this view is the cynical belief that humans are entirely self-seeking creatures unable or unwilling to sacrifice for the common good, especially if that good is some time off in the future. In short, we are presumed to be consumers with desires, not citizens, parents, neighbors, and friends with duties. They propose, accordingly, that in the shaping of environmental policy "governments [ought] to be more cynical about human nature" (ibid., 86), which is to say, government must buy off its citizenry.

Aside from the fact that such views tend to promote the very behavior they purport only to describe, what's wrong here? For one thing, the view does not square with the evidence from the grass roots, where outraged citizens attend rallies, march, and organize to stop the dams, highways, toxic waste dumps, clear-cuts, and shopping malls proposed by the rational self-maximizers. Not a few risk a great deal to do so. Why? Precisely because they are fed up with cynicism and greed and are willing to sacrifice a great deal for their communities, their children's future, and for a vision of something better. Furthermore, imagine for a moment Winston Churchill instead of saying to the British people in 1940, "I have nothing to offer but blood, toil, tears and sweat," saying something like "I'd like to ask each one of you to check your stock portfolios, bank accounts, and personal desires and if you are so inclined let us know what you are willing to do." A deal with Adolf Hitler would have been promptly struck. The fact is that we face a global emergency for which self-interest alone is woefully inadequate in the absence of deeper attachments and loyalties. To bring the enormous and destructive momentum of the human enterprise to a sustainable condition will require much more of us than the exercise of our individual self-interest would have us do, the kinds of things we are moved to do, in William James' words, because of "the big fears, loves, and indignations; or else the deeply penetrating appeal of some one of the higher fidelities, like justice, truth, or freedom" (James 1955, 211).

Rational self-interest, furthermore, seldom generates much imagination, creativity, and foresight, which will be greatly needed in coming decades. Philosopher Mary Midgley puts it this way: "Narrowly selfish people tend not to be very imaginative, and often fail to look far ahead. . . . Exclusive self-interest tends by its very nature *not* to be enlightened, because the imagination which has shrunk so far as to exclude consideration for one's neighbors also becomes weakened in its power to foresee future changes" (1985, 143). The reason that rational calculation alone does not amount to much has to do with how the embodied mind actually works. In the words of neuroscientist Antonio Damasio, "New neurological evidence suggests that . . . emotion may well be the support system without which the edifice of reason cannot function properly and may even collapse" (1994, 144). Emotion, far from being antithetical to rational thought, is a prerequisite for it.

The crisis of sustainability is nothing less than a test of our total character as a civilization and of our "personal aptitude or incapacity for moral life" (James 1955, 214). That being so, putting twine in the baler will mean expanding our perception of self-interest to include our membership in the larger enterprise of life over a longer sweep of time, and doing so with all the emotionally driven rationality we can muster.

Conclusion

Institutions purportedly dedicated to the life of the mind often suffer their own peculiar version of the twineless baler problem. Ideally, however, no institutions in modern society are better situated and none more obliged to facilitate the transition to a sustainable future than colleges and universities. If the public dialogue about sustainability gets beyond symbolism and down to hard realities, it will be because a much more fully educated and morally energized citizenry demanded it. What would it mean for educational institutions to meet this challenge?

For one thing, it would mean fostering, in every way possible, a broad and ongoing dialogue about concentrated economic power and the changes that will be necessary to build a sustainable economy. I know of no safe way to conduct that conversation that would not threaten the comfortable or risk losing some of the institution's financial support, a sensitive topic when the average cost of a college education is becoming prohibitively expensive.

Furthermore, colleges and universities ought to equip students, by every means possible, to think systematically, rationally, and, yes, emotionally about long-term technological choices and how such decisions ought to be made. That discussion, too, would raise contentious issues having to do with the meaning of progress and economic growth. And it would implicitly challenge the unbridled freedom of inquiry, if the extreme exercise of that freedom undermines biological order, democratic institutions, and social stability that gave rise to it in the first place. Issues of "who gains and who loses from unrestricted inquiry will press heavily on the university" (Michael 1993, 201) and cannot be dodged much longer.

Finally, the cynical view, pawned off as "objective" social science, that humans are only self-maximizers must be revealed for what it is: half-truth in service to the economy of greed. Increasingly, the young know that their inheritance is being spent carelessly and sometimes fraudulently. I believe that a sizable number know in their bones the truth of Goethe's words that "whatever you can do or dream you can, begin it, boldness has genius, power, and magic in it." What they may not know is where we, their teachers, mentors, and role models stand or what we stand for.

11

Conservation and Conservatism

The philosophy of free-market conservatism has swept the political field virtually everywhere, and virtually everywhere conservatives have been, in varying degrees, hostile to the cause of conservation. This is a problem of great consequence for the long-term human prospect because of the sheer political power of conservative governments. Conservatism and conservation share more than a common linguistic heritage. Consistently applied they are, in fact, natural allies. To make such a case, however, it is necessary first to say what conservatism is.

Conservative philosopher Russell Kirk (1982, xv–xvii) proposes six "first principles" of conservatism. Accordingly, true conservatives:

- believe in a transcendent moral order
- prefer social continuity (i.e., the "devil they know to the devil they don't know")
- believe in "the wisdom of our ancestors"
- are guided by prudence

- "feel affection for the proliferating intricacy of long-established social institutions"
- believe that "human nature suffers irremediably from certain faults."

For Kirk the essence of conservatism is the "love of order" (1982, xxxvi). Eighteenth-century British philosopher and statesman Edmund Burke, the founding father of modern conservatism and as much admired as he is unread, defined the goal of order more specifically as one which harmonized the distant past with the distant future. To this end Burke thought in terms of a contract, but not one about "things subservient only to the gross animal existence of a temporary and perishable nature." Burke's societal contract was not, in other words, about tax breaks for those who don't need them, but about a partnership promoting science, art, virtue, and perfection, none of which could be achieved by a single generation without veneration for the past and a healthy regard for those to follow. Burke's contract, therefore, was between "those who are living, those who are dead, and those who are to be born . . . linking the lower with the higher natures, connecting the visible and invisible world" ([1790] 1986, 194–195). The role of government, those "possessing any portion of power," in Burke's words, "ought to be strongly and awefully impressed with an idea that they act in trust" (ibid., 190). For Burke, liberty in this contractual state was "not solitary, unconnected, individual, selfish Liberty. As if every man was to regulate the whole of his conduct by his own will." Rather, he defined liberty as "social freedom. It is that state of things in which liberty is secured by the equality of restraint" (quoted in O'Brien 1992, 390).

As the ecological shadow of the present over future generations has lengthened, the wisdom of Burke's concern for the welfare of future generations has become more evident. Moreover, if conservatism means anything at all other than the preservation of the rules by which one class enriches itself at the expense of another, it means the conservation of what Burke called "an entailed inheritance derived to us from our forefathers, and to be transmitted to our posterity; as an estate belonging to the people" (Burke [1790] 1986, 119). Were Burke alive today, there can be no doubt that he would agree that this inheritance must include not only the laws, traditions, and customs of society, but also the ecological foundations on which law, tradition, custom, and public order inevitably depend. A society that will not conserve its topsoil cannot preserve social order for long. A society that wastes its natural heritage like a spendthrift heir can build only the most fleeting prosperity, leaving all who follow in perpetual misery. And those societies that disrupt the earth's biogeochemical balances and destroy its biota are the most radical of all. If not restrained, they could force all thereafter to live in an ecological ruin and impoverishment that we can scarcely imagine.

In light of Burke's view that society is a contract between the living, the dead, and those to be born, what can be said about the conservatism of contemporary conservatives? What, for instance, is conservative about conservatives' support for below market-cost grazing fees that federal agencies charge ranchers for their use of public lands? Welfare for ranchers runs against conservatives' supposed antipathy for handouts to anyone. But that's a quibble. The more serious issue concerns the ecological effects of overgrazing which result from underpricing the use of public lands. Throughout much of the American West, the damage to the ecology of fragile ecosystems is serious and increasing, with worse yet to come. In a matter of decades these trends will jeopardize a way of life and a ranching economy that can be sustained for future generations only by astute husbandry of soils, wildlife, and biota of arid regions. The ruin now being visited on a large part of public lands for a short-lived gain for a few is a breach of trust with the future. There is nothing whatsoever conservative about a system that helps those who do not need it while failing to sustain the ecological basis for a ranching economy into the distant future.

What is conservative about the ongoing support many conservatives give to the Mining Law of 1872? That piece of archaic legislative banditry permits the destruction and looting of public lands in the service of private greed while requiring little or nothing in return. The result—economic profligacy and ecological ruin—meets no conceivable test of genuinely conservative ideals and philosophy. It is theft on a grand scale, permitted because of the political power of those doing the looting and the cowardice and shortsightedness of those doing the governing.

What is conservative about getting government off the backs of citizens while leaving corporations there? Burke, who had a healthy dislike for all abuses of power, would have wanted all tyranny curtailed, including that of corporations. How do price increases, for example, differ from tax increases? How do cancers caused by toxic emissions or deaths resulting from safety defects in automobiles differ from unjust executions? How does the ability of capital to abandon one community for another that it can exploit more thoroughly differ from government mismanagement? To those who suffer the consequences, such differences are largely academic. The point is lost, nonetheless, on most contemporary conservatives who often detect the sins of government in parts-per-billion while overlooking corporate malfeasance by the ton. Burke, in our time, would not have been so negligent about economic tyranny.

What is conservative about squandering for all time our biological heritage under the guise of protecting temporary property rights? Conservatives have long scorned public efforts, meager as they are, to protect endangered species because, on occasion, doing so may infringe on the ability of property owners to enrich themselves. Any restrictions on private property use, even those which are beneficial to the public and in the interest of posterity, they regard as an unlawful taking of property. But this view of property rights finds little defense in a careful reading of either John Locke, from whom we've derived much of our land-use law and philosophy (Caldwell and Shrader-Frechette 1993), or in the writings of Burke. For Locke, property rights were valid only as long as they did not infringe on the rights of others to have "enough and as good" ([1690] 1963, 329). It is reasonable to believe that this ought to include the rights of future generations to a biota as abundant and as good as that which sustained earlier generations. And for Locke, "nothing was made by God for Man to spoil or destroy" (ibid., 332), a concept that has not yet been fully noted by many conservatives. The point is that Locke did not regard property rights as absolute even in a world with a total population of less than 1 billion, and neither should we in a world of 6.3 billion and rising.

What's conservative about a quarter century of opposition to national efforts to promote energy and resource efficiency? Even on narrow economic grounds, efficiency has been shown to be economically advantageous. The fact that the United States is far less efficient in its use of energy than Japan and Germany, for instance, places it at a competitive disadvantage estimated to be between 5 and 8 percent for comparable goods and services. Economics aside, energy and resource profligacy is the driving force behind climatic change and the sharp decline in biological diversity worldwide. Nothing could be more deleterious to the interests of future generations than for this generation to leave behind an unstable climate and the possibility that those changes might be rapid and self-reinforcing. And short of nuclear war, no act by the present generation would constitute a greater dereliction of duty or breech of trust with its descendants.

The willingness of many conservatives to accept the risk of catastrophic and irreversible global changes that would undermine the well-being of future generations is a profoundly imprudent precedent. We have no right to run such risks when the consequences will fall most heavily on those who can have no part in making the choice.

What is conservative about the extension of market philosophy and narrow economic standards into the realm of public policy? Many conservatives want to make government work just like business works. Government certainly ought to do its work efficiently, often much more efficiently than it now does. That much is common sense, but it is a far cry from believing that public affairs can be conducted as a business or that economic efficiency alone is an adequate substitute for farsighted public policy. Many good things, including compassion, justice, human dignity, environmental quality, the preservation of natural areas and wildlife, art, poetry, music, libraries, stable communities, education, and public spiritedness can never meet a narrow test of profitability, nor should they be required to do so. This, too, is common sense. These things are good in and of themselves and should not be subject to the same standards used for selling beer and automobiles.

What is conservative about perpetual economic growth? Economic expansion has become the most radicalizing force for change in the modern world. Given enough time, it will first cheapen and then destroy the legacy we pass on to the future. The ecological results of economic growth at its present scale and velocity are pollution, resource exhaustion, climatic instability, and biotic impoverishment. Uncontrolled economic growth destroys communities, traditions, and cultural diversity. And through the sophisticated cultivation of the seven deadly sins of pride, envy, anger, sloth, avarice, gluttony, and lust, economic growth destroys the character and virtues of the people whose wants it purports to satisfy.

Conservatives (and liberals) have been unwilling to confront the difference between growth and real prosperity and to tally up the full costs of growth for our descendants. In the words of former Reagan administration Defense Department official Fred Ikle, "Growth utopianism is a gigantic global Ponzi scheme [leading to] collapse, engulfing everyone one in misery" (1994, 44). Ikle continues to say that the cause of this collapse would not be a shortage of material goods but the destruction of society's conservative conscience by our Jacobins of growth.

That conservatives, by and large, have been deeply hostile to evidence of ecological deterioration and to the cause of conservation is profoundly unconservative. A genuine and consistent conservatism would aim to conserve the biological and ecological foundations of social order and pass both on as part of "an entailed inheritance derived to us from our forefathers and to be transmitted to our posterity" (Burke [1790] 1986, 119). If words mean anything, there can be no other standard for an authentic conservatism.

Like that defined in Kirk's first principles, a genuine conservatism is grounded in the belief in a transcendent moral order in which our proper role is that of trustees subject to higher authority. It would honor and respect the need for both social and ecological continuity. It would respect the wisdom of past and also the biological wisdom contained in the past millions of years of evolution. A genuine conservatism would prudently avoid jeopardizing our legacy to future generations for any reason of temporary economic advantage. It would eschew cultural and technological homogeneity and conserve diversity of all kinds. And a genuine conservatism, chastened by the recognition of human imperfection, would not create technological, economic, and social conditions in which imperfect and ignorant humans might wreak ecological havoc.

An authentic conservatism has much to offer in the cause of conservation. Conservatives are right that markets, under some circumstances, can be more effective tools for conservation than government regulation. The conservative dislike of unwarranted taxation might be the basis on which to shift taxes from things we want, such as income, profit, and labor, to things we do not want, such as pollution and energy and resource inefficiency (von Weiszacker and Jesinghaus, 1994). An authentic conservatism would encourage a sense of discipline, frugality, and thrift in the recognition that "men are qualified for civil liberty in exact proportion to their disposition to put moral chains upon their own appetites. . . . Society cannot exist unless a controlling power upon will and appetite be placed somewhere, and the less of it there is within, the more there must be without. It is ordained in the eternal constitution of things, that men of intemperate minds cannot be free. Their passions forge their fetters" (Burke, quoted in epigraph to Ophuls 1992). A genuine conservatism would provide the philosophical bases and political arguments for prudence, precaution, and prevention in public policy and law. And a genuine conservatism would recognize that avoidance of some tragedies requires "mutual coercion, mutually agreed upon" (Hardin 1968, 12), which, in turn, requires robust democratic institutions.

12

A Politics Worthy of the Name

> Genuine politics—politics worthy of the name . . . is simply a
> matter of serving those around us: serving the community, and
> serving those who will come after us.
>
> —*Vaclav Havel*

Relative to the problems we face, our politics are about the most mis-
erable that can be imagined. Those who purport to represent us and
who on rare occasions try to lead us have been unable to take even
the smallest steps to promote energy efficiency to avoid possibly catastrophic climatic change a few decades from now. They have failed
to stop the hemorrhaging of life and protect biological diversity, soils,
and forests. They ignore problems of urban decay, suburban sprawl,
the poisoning of our children by persistent toxins, the destruction of
rural communities, and the growing disparity between the rich and
the poor. They cannot find the wherewithal to defend the public interest in matters of global trade or even in the financing of public elec-

tions. Indeed, the more potentially catastrophic the issue, the less likely it is to receive serious and sustained attention from political leaders at any level.

Our public priorities, in other words, are upside down. Issues that will seem trivial or even nonsensical to our progeny are given great attention, while problems crucial to their well-being are ignored and allowed to grow into global catastrophes. At best they will regard us with pity, at worst as derelict and perhaps criminally so. The situation was not always this way. The leadership of this country was once capable of responding to threats to our security and health with alacrity and sometimes with intelligence.

In light of the dismal performance of the U.S. political system relative to the large environmental and social issues looming ahead, we have, broadly speaking, three possible courses of action (assuming that we choose to act). The first is to turn the management of our environmental affairs over to a kind of permanent technocracy—a priesthood of global managers. The idea that experts ought to manage public affairs is at least as old as Plato. In its current incarnation, some propose to turn the management of the earth over to a group of global experts. Stripped to its essentials, this means smarter exploitation of nature culminating in the global administration of the planet with lots of satellites, remote sensing, and geographic information systems experts mapping one thing or another. The goal of smarter ecological management is to keep the extractive economy going a bit longer by merely improving our management instead of rethinking our aims (Sachs 1999). Technocrats will manage the environment efficiently without much public participation or discussion of goals. If history is any indication, they will ride roughshod over communities, indigenous people, native cultures, farmers, and small landowners. Planet managers will hold expensive conferences in exotic places, issue glossy and reassuring reports, and ingratiate themselves with the rich and powerful. In the end, however, they will fail because the knowledge, foresight, and wisdom necessary for planetary management are beyond human grasp and because people everywhere will reject imperialism in its new guise of planetary management.

A second possibility is to admit that all politics is really about economics anyway and turn things over to business corporations and the market. Given the scale of our problems, the need for quick action, and the difficulties of reforming democracy, there is much to be said for turning matters over to people who know how to get things done. But capitalism, whatever its other qualities, is not famous for protecting environments or serving the public interest. Could it be reformed along ecological lines? Some believe so. Factories would be made over into industrial ecologies in which every waste product would be used somewhere else. Businesses would sell "products of service," not just consumer goods, that are forever turned back into new product. They would sell green and energy-efficient products. Taxes would be levied on things we do not want such as pollution and removed from those that we do want such as income and profits. Above all, an ecologically solvent capitalism would account for its environmental and social costs.

An ecologically reformed capitalism would be a great improvement on the present system. As a strategy of change it is logical because capitalism is virtually everywhere ascendant and governments everywhere seem to be in retreat. Business, in short, is where the action is. Operating along the model of ecosystems, businesses presumably would not require close regulation. The role of government, therefore, would be minimal and the need for a democratically informed citizenry would diminish accordingly. Best of all, relying on business to lead the transformation would require little of the public. Instead, the logic of enlightened economic self-interest would drive us toward a sustainable relationship with nature. But why would capitalism, a system based on ruthless pursuit of short-term self-interest, yield to such changes? If it were only a matter of logic, a decent concern for our grandchildren, or even enlightened self-interest, we could be optimistic, but alas, the issue is not so simple.

First, there is the question of whether it is possible to redesign capitalism to accord with ecological realities. The problem is simply that "the self-organizing principles of markets that have emerged in human cultures over the past 10,000 years are inherently in conflict with the self-organizing principles of ecosystems that have evolved over the past 3.5 billion years" (Gowdy and McDaniel 1995, 181). Markets are inappropriate tools to solve many problems of ecological scarcity. For example, blue-fin tuna have been fished almost to extinction. But the logic of the unrestrained market will not reduce the take but, rather, will work to ensure that the last blue-fin tuna, selling for hundreds of thousands of dollars, will be caught and sold and the money invested elsewhere. The owners of capital do not care whether they make money in fisheries or condominiums. The logic of exploitation is relentless, predisposing the system to tragic ends with many luxury goods but few fish.

The problem, in other words, is not that capitalists lack the right information about the full ecological costs of what they do, but rather that capitalism and ecological management are two fundamentally different value systems that aim at different things. Markets, driven by the logic of self-interest, are intended to maximize profits and minimize costs for the owners of capital in the short term. Ecosystems, in contrast, operate by the laws of thermodynamics and processes of evolution and ecology that are played out over the long term.

Second, the possibility that increasingly powerful and predatory corporations will reform themselves is remote while countervailing forces, governments, an active citizenry, and labor unions are in decline. The political arrangements of the New Deal that tamed some of the worst excesses of U.S. capitalism for a time have come undone. Now a global capitalism in the age of free trade is more powerful and less restrained than ever. The result is a kind of robber baron phase of global economic history with no remedy in sight (Soros 1997). Corporations now operating in a free-trade environment have fewer constraints than ever before. The problem is compounded by the several trillion dollars that wash around the planet each day in search of the highest rates of return. The results of footloose capital and unrestrained corporate power are all too clear: too many dams, too many cars, too many shopping malls, too many mines, too many factories, and toothless environmental controls.

Third, the discipline of economics that explains, informs, and justifies capitalism and educates capitalists has so far successfully resisted accommodation with ecology and thermodynamics. The profession has proven to be largely impervious to the devastating critiques of maverick economists such as Kenneth Boulding, Nicholas Georgescu-Roegen, Herman Daly, Robert Constanza, John Gowdy, and Hazel Henderson. Logic, data, and evidence, notwithstanding, mainstream economists hold with remarkable tenacity to beliefs that technology can substitute for the loss of natural capital, economies can grow without limits, and human desires are insatiable. Both the profession of economics and its practice as capitalism are perpetuated as belief systems by denial, repression, alienation from life, addiction, and what theologian Thomas Berry (1999) calls a kind of ecological autism (see also Gladwin et al. 1997). The collective irrationality masquerading as realism or even science, in other words, is a manifestation of life-denying pathologies that are now deeply embedded in a professional caste.

Fourth, a reformed capitalism is still capitalism—a system that thrives only when people buy and buy more than they need. Even if they make "green" products and recycle all of their wastes, corporations, for reasons of scale and power, will act to undermine political participation, weaken the sense of community, and subvert democracy. Even a reformed capitalism would still be a system that works best when people confuse who they are with what they own. And it would still be a system that must move large volumes of stuff long distances as rapidly as possible. Capitalism, once a system largely contained within national borders, has evolved into a global system in which consumers cannot know the larger human and ecological costs of the system that provisions them and in which sellers cannot be held accountable for what they do.

Capitalism, in other words, is no more likely to transform itself into ecotopia than lions are to become vegetarians. We urgently need an economy that works ecologically, but the decision to reform capitalism or to invent some other kind of economy is a political, not an economic, choice. Issues having to do with the distribution of costs, benefits, risks, and wealth within and between generations are matters of fairness and decency, not efficiency. The scale of the economy relative to the environment is a political choice that can be made only by an ecologically literate public. Capitalism on its own is expansive and will ride roughshod over boundaries and limits of all kinds. If limits are imposed on the economy, they must be imposed politically by a citizenry that knows when enough is enough. Questions of what to tax and how to distribute public revenues wisely have to do with justice, fairness, accountability, and ecological prudence. These are political decisions. The economy, in other words, is a means, not an end.

The third possibility—and our only real choice—is to create a better kind of politics and political institutions better suited to ecological realities. The task would require rethinking the foundations of public life much as the founders of this republic did in the eighteenth century. To do so we would have to rethink basic questions of political life as they did, but in recognition of ecological facts which they did not know. The challenge before us is a design problem: how to build a decent civilization that fits harmoniously into the ecology of North America over the long term.

We are not accustomed to thinking of the effects of political decisions in the long term, let alone as a problem of ecological design. In fact, we've come to think of politics as mostly having to do with jobs and economic growth in the short term. All of the ideologies of the twentieth century—capitalism, communism, socialism, and fascism—are essentially competing views about how to organize industrial society. For all of the wars and ideological huffing and puffing, the differences between them in historical perspective are quibbles having to do with who owned and managed capital. Otherwise agreement prevailed that humans ought to dominate nature, technology should be unfettered, that we should burn fossil fuels as rapidly as possible, and that economic growth is the supreme value. Politics was reduced to questions having to do with the ownership of the means of production and how to distribute the profits. Political views, accordingly, arrayed themselves along a single axis of left to right denoting the extent to which one favored public or private control of capital. But we have entered a new political era in which the Left/Right dichotomy no longer works, not because questions of ownership are unimportant but because other issues have surged to the forefront.

These issues were there all along, of course. In *The Great Frontier*, historian Walter Prescott Webb described the great increase in per capita wealth generated by the discovery of the New World. The ratios of people to land and resources were fundamentally transformed until the middle of the twentieth century, when they once again approximated those of the year 1500. The rapid exploitation of fossil fuels has allowed us to continue the expansion for a while longer, but the end of the human efflorescence has come into view. "The modern age," Webb wrote, "was an abnormal age. . . . The institutions developed in this exceptional period are exceptional institutions" (1964, 14). At the end of the boom those institutions "and their attendant ideas about human beings, government, and economics . . . may be expected to undergo much change when those conditions have passed away and history returns to normal" (ibid.).

James Madison had a premonition that we would come to such a time. Richard Matthews says of Madison: "A Malthusian before Malthus, he constructed a political system that would postpone the inevitable decay for as long as reason would allow" (1994, 244).

The inevitable for Madison would be caused by a surplus of consumers created by population growth and technological development. The Louisiana Purchase and continental expansion would buy some time but would not resolve the underlying political problems of eventual scarcity. Good Calvinist that he was, Madison sought only to delay what he regarded as inevitable, but he could see no way out (1994, 210).

The Great Frontier is now spent; we live on a full planet. There will be attempts to extend the boom a while longer by heroic technology such as genetic engineering. When they fail, we will have to rethink the foundations of political life, retracing the steps of Madison, Jefferson, Hamilton, and the other architects of modern politics but under much less favorable conditions and without the safety valve provided by the frontier. The end of the Great Frontier means, in short, that we can no longer avoid basic political issues of fair distribution of wealth within and between generations by expanding production to keep the poor content. Discarding old truisms about rising tides lifting all boats and larger pies, we will be forced to reconsider politics and economics relative to the limits of the biosphere and in relation to the way the world works as a physical system.

In this light, societies have only four choices about how they provision themselves with food, energy, materials, and water, and how they dispose of their wastes. The choices have to do with

- how far the things used or consumed are transported
- the rate at which materials are used up and discarded
- the volume of materials used
- the sources of energy that power the entire system.

Until the industrial revolution, all societies met their basic needs locally or regionally. The rate and volume of resource use was low, and populations grew slowly if at all. Energy was derived from contemporary sunlight in its various forms of biomass, wind, and water power.

In contrast, we are supplied by a global network of forests, farms, mines, wells, and factories powered by the combustion of large amounts of fossil fuels. Population growth is high. We measure our success in terms of the gross national product, which is roughly the speed and volume with which materials flow through the economic pipeline from mines, wells, forests, and farms to dumps, smokestacks, and outfall pipes. In the language of physics, this is the rate at which we convert ordered matter or low entropy into waste and heat or high entropy. To keep this system going we provide easy and underpriced access to resources and lucrative tax and financial incentives to extractive industries and subsidize timber cutting, road building, automobiles, energy generation, and land sprawl (Myers 1998). And to keep demand growing, corporations spend perhaps as much as $500 billion each year on advertising (United Nations 1998, 7). Environmental protection is an add-on in the form of pollution control at the end of the entropic pipeline and comes too late in the process to be effective.

The large-scale systems and global organizations established to provide us with an abundance of cheap food, fossil energy, materials, and water and dispose of our wastes were created on assumptions that nature was inexhaustible and that human actions counted for little given the immense bounty of nature. At a scale far greater than their creators could have imagined, those systems have nearly ruined us. They have degraded our landscapes and ecosystems, spread toxins worldwide, weakened community ties, undermined our democracy, and reduced our capacity to take responsibility for what we do because we cannot know what we are doing or undoing. These are not side effects or accidents but predictable results of the way we have organized the flow of food, materials, energy, and water.

We take great pride, for example, in being the best, and most cheaply, fed people in history. But we are fed by a ruinous fossil fuel–powered industrial system that contributes to climatic change, water pollution, biotic impoverishment, depletion of groundwater, and soil loss. It exploits labor and rural communities and undermines future productivity of the land. The system encourages obesity, cancer, and heart disease—all signs of a national eating disorder. Given its scale and complexity, it cannot work responsibly, nor can consumers, ignorant of how it works, know enough to eat responsibly. The system dominated by large agribusiness firms, petrochemical companies, and seed companies undermines democracy. In fact, it works only to the extent that real democracy does not work and people do not know these things or do not see them as part of a larger pattern or fail to see opportunities to create a better food system.

These problems are not isolated events or accidents in an otherwise good system. They are, rather, the logical results of a bad system that just grew without anyone thinking much about how it fit (or did not fit) into the patterns set by ecology, evolution, thermodynamics, community, or democracy. If we want a better politics, we must first design better ways to meet our essential needs and remove the sources of tyranny from our lives. To do so we must take greater responsibility for how we are fed and supplied, replacing the elaborately destructive systems that provision us with better ones that rely on local resources and local competence. We cannot make democracy work unless we can make it work with, not against, the ecology of the particular places in which we live. By whatever name, the alternatives to large-scale, corporate control of our lives and politics require that people, neighborhoods, and communities assume a larger responsibility for meeting their own needs. The roots go back to Thomas Jefferson. The enemy in his time and ours is what he termed "remote tyranny." For Jefferson that meant the king and Parliament living an ocean away. In our time remote tyranny means both geographically remote and remote in time—in other words, any source of unaccountable power, corporate, governmental, or societal.

There is no way to hold a global economy accountable. Consequently, people and local communities are defenseless, without any good way to redress grievances or protect themselves from crises elsewhere. In a global system, a crisis anywhere becomes a crisis everywhere. There is no buffer, no margin, and no recourse when things go bust. It is now possible to see that Jefferson, for all of his ambiguities, was the great realist and Alexander Hamilton the dreamer. Jefferson knew what Hamilton and his followers did not know: that the health of democracy and that of the economy can be maintained only if citizens control the basic circumstances of their lives and livelihood. Jefferson's alternative plan stressed local independence, agrarianism, public accountability, widespread land ownership, and democratic participation. Hamilton's vision prevailed, at least for a time, but Jefferson's retains a hold on the human imagination virtually everywhere. Vaclav Havel, for example, describes a Jeffersonian vision for the Czech Republic in these words:

> Every main street will have at least two bakeries, two sweet-shops, two pubs, and many other small shops, all privately owned and independent. . . . Small communities will naturally begin to form again, communities centred on the street,

the apartment block, or the neighbourhood. People will once more begin to experience the phenomenon of home. It will no longer be possible, as it has been, for people not to know what town they find themselves in because everything looks the same. . . . Our villages will once again have become villages. . . . Agriculture should once again be in the hands of the farmers—people who own the land, the meadows, the orchards, and the livestock, and take care of them. In part, these will be small farmers who have been given back what was taken from them. . . . A pluralistic network of processing and marketing cooperatives, to which farmers belong, will exist. (Havel 1992, 104, 110–112)

I would tend to favor an economic system based on maximum possible plurality of many decentralized, structurally varied, and preferably small enterprises that respect the specific nature of different localities and different traditions and that resist the pressures of uniformity by maintaining a plurality of modes of ownership and economic decision-making from private through various types of cooperative and shareholding ventures, collective ownerships. (Havel 1991, 16)

Jefferson's vision for this country was never really tried. Instead, it was dismissed in the national rush to expand to continental proportions and to become a world power. Even though it is dismissed as impractical, it is still trotted out for sentimental reasons from time to time. But knowing more about the ecological and human costs of Hamilton's vision of America, Jefferson's looks better and better with the passage of time. So, too, does his idea that no generation ought to impose debt on succeeding ones. In a famous letter to James Madison in 1789, Jefferson asked whether "one generation of men has a right to bind another." His answer was no based on the principle that "the earth belongs to the living and not to the dead." Jefferson concluded, "No generation can contract debts greater than may be paid during the course of it's own existence" (1975, 244).[1] Were he alive now,

---

1. Madison's initial response was not positive. He objected that some debt incurred for "improvements" or "repelling conquest" benefited posterity.

I think that Jefferson would agree that the dead could also encumber the living by leaving behind depleted soils, denuded landscapes, hazardous wastes, biotic impoverishment, and changing climate; debt could be both ecological and financial.

For us, Jefferson's political vision has two great advantages. First, his insistence that no generation encumber the future with debt is a principle that transcends the present impasse between liberals and conservatives and bears resemblance to the views of Edmund Burke described in chapter 11. Jefferson, a man of the Left, and Burke, the patron saint of modern conservatism, both agreed that decisions of the present must be measured against the degree to which they encumbered future generations. Both saw the possibility that tyranny might be remote in time as well as in space. Writing within a year of each other, the views of the founders of modern conservatism and modern radicalism converged on a similar point: the welfare of future generations. That standard cuts across the divisions between Left and Right that have stalled our national politics. It coincides with every major religion in the world, and it appeals to the heart as well as to practical reason.

The second great virtue of Jefferson's vision is that it coincides with what we have come to understand as the principles of resilient systems that can withstand outside disturbances. Principles derived from ecology, systems theory, engineering, mathematics, and the study of the evolution of living systems over 3.8 billion years bear a strong similarity to those Jefferson proposed for the new nation. The basic design principles for resilient systems of all kinds have common characteristics (Lovins and Lehmann 1977, Lovins and Lovins 1982), such as:

- small units dispersed in space
- redundancy
- short linkages between modules
- simplicity and repairability
- diversity of components

Eventually, however, Madison came to accept the idea of limiting the public debt for reasons similar to those originally proposed by Jefferson (Matthews 1995).

- self-reliance
- decentralized control
- large margins
- quick feedback.

Jefferson's nation of small farmers no longer exists, but the underlying principles are still valid. For his time Jefferson proposed the creation of a society capable of preserving democracy while withstanding the turmoil of a simpler agrarian world. In the twenty-first century, that same goal would aim to create resilient communities that provide a large fraction of their own food, energy, shelter, health, recreation, and financing in order to withstand global financial crises, volatile stock markets, the effects of capital mobility, corporate downsizing, terrorism, and interruption of resource supplies. More resilient communities would create more of their own jobs without importing footloose capital. They would control most of their own money. Ownership would be widespread (Gates 1998, Shuman 1998). They would grow a large fraction of their own food locally or regionally. They would utilize local and renewable energy to the maximum. The sophisticated modern mind atrophied by all of the nonsense about the global economy and the necessity for economic growth has dismissed these notions as nostalgia or worse. In fact, resilience and democracy both require a social order that features rich community life, neighborliness, competence, self-reliance, human scale, and ecological durability.

We need not expect help from those who fatten at the trough of the global economy. The reason is simple: money—specifically the $125 billion in welfare handed out by the federal government to corporations, the $300 billion in subsidies for highways and automobiles, and the $1.4 trillion in global subsidies for environmental destruction (Barlett and Steele 1998, Myers 1998). Until such time as we have the good sense to establish a complete and total separation between money and politics—like that between church and state—our national and state politics will be corrupt and ineffective. We must remove money from politics at all levels once and for all. Federal funding for national elections is a start. The next step is to rein in the power of corporations by insisting that they abide by the terms of their charters. The charters of those that cannot do business within the terms of the law should be revoked. A corporate version of "three strikes and you're out," for instance, would have a salutary effect on corporate behavior.

None of this, however, is likely to begin in Washington, D.C. It will have to begin in communities, towns, urban neighborhoods where consumers decide to become citizens and take control of their lives and livelihood. The effect would be a diminution of power of those who cultivate what Jefferson called dependence and venality. Local food production and cooperatives would begin to weaken the power of the giant food monopolies. Power systems distributed to rooftops and buildings would weaken the hold of giant utilities. Local currencies and local investments would weaken the hold of financial speculators and money brokers. Every alternative to the consumption of gasoline, from better designed communities to cars that run on solar hydrogen, would weaken the hold of the giant oil companies. Over years and decades the quiet withdrawal from large-scale systems reduces the prospect of ecological catastrophe, social injustice, and remote tyranny. A more resilient social order does not guarantee the rejuvenation of democracy, but it does change what the public perceives to be possible. Every solar collector, every community garden or wind farm, every local currency is a declaration of independence from remote tyranny and a declaration of interdependence with all of life and with generations unborn. The eventual reform of national politics will begin when elites begin to feel the desperation that comes from the awareness of being left behind. The strategy is the same as that described by Lewis Mumford, who once proposed to use the power of "animated individual minds, small groups, and local communities" not to seize power, but to "withdraw from it and quietly paralyze it" (1970, 408).

You and I will have to do the hard work of reviving democracy and rebuilding a decent country and ecologically sustainable communities the old-fashioned way: from the bottom up. There is no use pretending that it will be easy to do, but it will be a great deal easier than vainly trying to make our peace with the forces of tyranny in our time. Eventually the small brigades will win for the same reasons that small mammals survived and the dinosaurs died out, that Drake's fleet defeated the Spanish Armada, and that all large organizations eventually become sclerotic and rigid. The reasons have to do with agility, the capacity to respond quickly, adaptability, and the principles of resilience. These are things that can be sustained only at an appropriate scale.

Eventually, urban neighborhoods, communities, small towns will quietly paralyze the sources of remote tyranny by withdrawing from them. The transformation, already under way, is easy to overlook because it is not dramatic, it does not make for good slogans, and it does not need a national organization. It is people taking back power by forming community-supported farms and land trusts, by using local currencies, by using less fossil fuels and more solar energy, by starting community businesses, and doing all of the hard work of becoming citizens again. The logic of decentralization—democracy from the bottom up—is founded on simple facts of how the world really does work. Local economies prosper by minimizing dependency on the outside economy and by meeting local needs with local resources.

Are we up to it? Time will tell. The sources of remote tyranny in our time prefer to keep us in a state of consumer-besotted ignorance. But, in Jefferson's (1816, 473) words, "If a nation expects to be ignorant and free . . . it expects what never was and never will be."

13

The Limits of Nature and the Educational Nature of Limits

I teach in a liberal arts college in a small, attractive Ohio town located in an agricultural county 14 miles south of Lake Erie. The town formerly had train service that connected it easily and comfortably to the wider world. Sometime in the 1950s the trains stopped coming, and the tracks were eventually converted into a bike trail. In the intervening four decades, students arrived on campus in a variety of ways, including bus, plane, car, and a few intrepid souls still come by train to a decaying Amtrak station eight miles distant. Now many, perhaps most, come in cars that they own and that they park anywhere and everywhere in town. So like many campuses, ours is overrun by cars. And like many other colleges, we find ourselves locked in conflict with the local authorities over parking policy. Our policy is roughly to tell students, "Y'all come and bring it with you."

Unless there is a sudden outbreak of intelligence, we are likely to respond to prodding by city officials by building yet another parking lot and thereby reducing to that degree the loveliness and serenity of the town already jeopardized by urban sprawl. That, however, is an aesthetic matter on which people can and will disagree. What they cannot dispute is the cost of parking. The cost of a single parking space is estimated to be $7,000 in a paved lot and double that for a parking deck. Then there is the annual cost of policing, lighting, removing snow, and landscaping parking lots, perhaps another $1,500. From this perspective, one obvious solution is simply not to build extra parking and split the savings with those who do not to bring cars to college or drive them to work. So in return for not adding to the problem, cooperators would get a check for, say, $5,000. Those who continue to drive for whatever reason would pay a fee equal to the real costs imposed on the institution by their driving habits. Reasonable? Not according to many who believe that driving is a sacred right guaranteed somewhere in the Constitution (or was it the Declaration of Independence?) and to those who believe that automobility is now indelibly written into our behavioral genes and cannot be further altered by evolution or reason.

This issue is instructive because it captures in a microcosm larger issues of scarcity and management of common problems. We now confront problems of scarcity in one form or another that can be solved only by some combination of smart incentives and, as Garrett Hardin once put it, "mutual coercion, mutually agreed upon" (1968, 12). Oberlin's parking problem is instructive, too, because it highlights the ways in which scarcity, for lack of a better phrase, is socially constructed. The town and the college are about the same size that they were 40 years ago. But our values, attitudes, and habits and consequently our perception of our possibilities, have changed. This issue is also instructive for what it says about our ability to solve problems in which technological fixes (parking lots) compete with social solutions (fees/rebates) and value change (walk or bike, don't drive). Finally, the manner in which the problem is resolved will either enhance or diminish our capacity to engage each other in a public dialogue and perhaps our level of civility as well. On a larger and more abstract level, much of the same is true as well.

In a widely influential article, for example, Mark Sagoff asserts that "it is simply wrong to believe that nature sets physical limits to economic growth" (1997, 83). Such opinions, he argues, hinge on the mistaken beliefs that mineral resources are finite, that we are running out of food and timber, that we are running out of energy, and that resource consumption by the wealthy North exploits the poorer nations of the Southern Hemisphere. Sagoff's view, as I understand it, is not necessarily that our present course can be sustained, but rather that better technology will help us surmount natural limits without requiring substantial changes beyond what we are willing to adopt. He, like the late Julian Simon, places a great deal of faith in human ingenuity. Sagoff, however, is no unregenerate hedonist advocating higher levels of consumption. On the contrary, he believes that there are indeed limits to resource use and consumption, but such limits are inherent in our spiritual needs for affiliation with nature, not in nature itself. "An intimacy with nature," he writes, "ends our isolation in the world" (Sagoff 1997, 96). His conclusion is simply that "the question before us is not whether we are going to run out of resources. It is whether economics is the appropriate context for thinking about environmental policy" (ibid., 96). The answer for Sagoff is a resounding no.

Predictably, Sagoff's article aroused vigorous dissent. Within the year, Paul Ehrlich and coauthors responded in the same forum that Sagoff "has done a disservice to the public by promoting once again the dangerous idea that technological fixes will solve the human predicament" (1997, 98). Their argument is roughly the inverse of Sagoff's: resources are indeed finite, nature's services are increasingly threatened by consumption, prices do not provide reliable signals of resource scarcity, technology is no magic solution, and, yes, the wealthy nations do exploit the resources and people of poorer nations.

Both positions are, of course, much more detailed than my brief synopsis suggests. On balance, however, the issues are familiar ones, dating back at least to the controversy over *The Limits to Growth* (Meadows et al. 1972). In the intervening years, the stakes have grown higher. Evidence mounts that humans are now impacting the global environment and eroding what has come to be called natural capital. Despite the emergence of a worldwide environmental movement, capitalism and consumerism are virtually everywhere triumphant. The global economy is reshaping the world for more consumption, not less. And not least, mass advertising aims to reshape the minds of young people to believe that consumption is their natural right. Corporate involvement in education at all levels is intended to create a generation of pliable minds incapable of thinking that consumption is anything other than natural or that the corporations that make it possible, fun, and convenient are anything other than friendly. These are minds that will come to regard economics as more basic than politics or ethics and that will view whatever problems we encounter as simply technological problems, not fundamental dilemmas. So before we pass some point of no return and discover that we are like bugs mashed on the windshield of illusions and error, we had better get the issue of limits, both natural and human, right.

Whether allocation of space at Oberlin College or management of the global commons, how are we to think about the limits of nature and those of society? First, there are few technological responses to limits that will not entail one ambiguity or another. Artifacts, as Langdon Winner (1980) once noted, have politics. Whether parking lots or genetically engineered agriculture, technological solutions rearrange our minds and our social, economic, and political institutions as well. Often they do so in ways that create unforeseen, deleterious, and irreversible outcomes, what Eugene Schwartz (1917) once identified as secondary and tertiary problems of technology. Solutions, as someone once put it, cause problems. And having rearranged our minds and politics, technologies create expectations and eventually entire constituencies that come to believe that the resulting unsustainable condition is the normal state. Political and economic power follows (Ludwig et al. 1993). All of this is to say that it is possible to respond to limits in ways that set in motion a chain of responses that, over time, diminish our flexibility and capacities to deal with still other and more strenuous limits.

Once we've paved over a large part of Oberlin, we not only will have promoted the very habits that require still more paving, but we will have diminished the funds and space necessary for, say, bike trails. Once having industrialized and engineered our entire food system, we will have lost much of the cultural information necessary to farm and feed ourselves in less precarious and more desirable ways. Technological solutions are not neutral. They skew power and resources in one way or another, affecting our abilities to deal with future limits in ways consonant with other values that we hold dear.

The belief that there are no limits to nature that cannot be stretched or eliminated by technology masquerades as the realistic view of things. In fact, such views rest on a kind of faith that closely resembles religious belief but without the heart and soul of authentic religion. In David Noble's words, "The religion of technology has become the common enchantment, not only of the designers of technology, but also of those caught up in, and undone by, their godly designs. The expectation of ultimate salvation through technology, whatever the immediate human and social costs, has become the unspoken orthodoxy, reinforced by a market-induced enthusiasm for novelty as sanctioned by a millenarian yearning for new beginnings" (1998, 207). And like all fundamentalists, adherents to the religion of technology regard "any and all criticism [as] irrelevant and irreverent" (ibid., 207). The result is often a pattern of denial that categorically dismisses the very concept of limits. And having dismissed the concept of limits, we will simply not see them when they present themselves to us, especially if they are the small things in nature or if they involve the slow loss of natural services. We will have lost the ability and patience to pay attention. Technology can extend our sight into the far reaches of space while reducing our ability to see what is before our very eyes.

Sagoff, having faith in our godlike ability to surmount natural constraints, proposes that we nonetheless limit ourselves to promote "affection and reverence for the natural world" (1997, 96). Nothing seems less likely than the idea that people, having perceived themselves to be beyond the limits of nature, would voluntarily limit their appetites for ostensibly spiritual or moral reasons. The opposite seems far more plausible. I think Sagoff has mistaken not only the limits of nature but the role that the awareness of limits plays in human psychology. We need some limits because they free us. It is the awareness of our finiteness that causes us to reckon with what's really important in life. The awareness of limits opens us to the fact of our unlimited dependence on a larger order of things that we will never fully comprehend. Gratitude and wonder, not technological escapism, are the appropriate responses. Real moral growth, I think, is built on the awareness of our limitations and the existence of larger limits that lead us to share and understand that the gift must move.

Finally, problems posed by limits, whether parking lots or the management of global carbon dioxide emissions, represent opportunities for civic education. Discussions about the Kyoto Protocol, for example, should not be confined to legislatures and government halls around the planet. Such agreements ought to be debated in every city, town, and village in the world. Only in this way will we come to regard self-governance as part of the living and common heritage of humankind. Similarly, the issue of parking, here and elsewhere, is an opportunity to educate communities about the limits of space, fairness, natural beauty, full-cost economics, the role of the automobile in society, tragedies of common property resources, and quite possibly, creative ways to solve common problems. It is also an opportunity to debate what kinds of communities we want to create and get on with the job of building them. What better educational opportunity could there be?

§ 4

DESIGN AS PEDAGOGY

14

Architecture and Education

> The worst thing we can do to our children is to convince them
> that ugliness is normal.
>
> —*Rene Dubos*

As commonly practiced, education has little to do with its specific
setting or locality. The typical campus is regarded mostly as a place
where learning occurs, but is, itself, believed to be the source of
no useful learning. A campus is intended, rather, to be convenient,
efficient, or aesthetically pleasing, but not instructional. It neither
requires nor facilitates competence or mindfulness. By that standard, the same education could happen as well in California or in
Kazakhstan, or on Mars, for that matter. The same could be said of
the buildings and landscape that make up a college campus (Orr
1993). The design of buildings and landscape is thought to have
little or nothing to do with the process of learning or the quality of
scholarship that occurs in a particular place. But in fact, buildings

and landscape reflect a hidden curriculum that powerfully influences the learning process.

The curriculum embedded in any building instructs as fully and as powerfully as any course taught in it. Most of my classes, for example, were once taught in a building that I think Descartes would have liked. It is a building with lots of squareness and straight lines. There is nothing whatsoever that reflects its locality in northeast Ohio in what had once been a vast forested wetland (Sherman 1996). How it is cooled, heated, and lighted and at what true cost to the world is an utter mystery to its occupants. It offers no clue about the origins of the materials used to build it. It tells no story. With only minor modifications it could be converted to use as a factory or prison, and some students are inclined to believe that it so functions. When classes are over, students seldom linger for long. The building resonates with no part of our biology, evolutionary experience, or aesthetic sensibilities. It reflects no understanding of ecology or ecological processes. It is intended to be functional, efficient, minimally offensive, and little more. But what else does it do?

First, it tells its users that locality, knowing where you are, is unimportant. To be sure, this is not said in so many words anywhere in this or any other building. Rather, it is said tacitly throughout the entire structure. Second, because it uses energy wastefully, the building tells its users that energy is cheap and abundant and can be squandered with no thought for the morrow. Third, nowhere in the building do students learn about the materials used in its construction or who was downwind or downstream from the wells, mines, forests, and manufacturing facilities where those materials originated or where they eventually will be discarded. And the lesson learned is mindlessness, which is to say, it teaches that disconnectedness is normal. And try as one might to teach that we are implicated in the larger enterprise of life, standard architectural design mostly conveys other lessons. There is often a miscalibration between what is taught in classes and the way buildings actually work. Buildings are provisioned with energy, materials, and water, and dispose of their waste in ways that say to students that the world is linear and that we are no part of the larger web of life. Finally, there is no apparent connection in this or any other building on campus to the larger set of issues having to do with climatic change, biotic impoverishment, and the unraveling of the fabric of life on earth. Students begin to suspect, I think, that those issues are unreal or that they are unsolvable in any practical way, or that they occur somewhere else.

Is it possible to design buildings and entire campuses in ways that promote ecological competence and mindfulness (Lyle 1994)? Through better design, is it possible to teach our students that our problems are solvable and that we are connected to the larger community of life? As an experiment, I organized a class of students in 1992–1993 to develop what architects call a preprogram for an environmental studies center at Oberlin College. Twenty-five students and a dozen architects met over two semesters to develop the core ideas for the project. The first order of business was to question why we ought to do anything at all. Once the need for facilities was established, the participants questioned whether we ought to build new facilities or renovate an existing building. Students and faculty examined possibilities to renovate an existing building, but decided on new construction. The basic program that emerged from the year-long class called for a 14,000-square-foot building that

- discharged no wastewater (i.e. drinking water in, drinking water out)
- eventually generated more electricity than it used
- used no materials known to be carcinogenic, mutagenic, or endocrine disrupting
- used energy and materials efficiently
- promoted competence with environmental technologies
- used products and materials grown or manufactured sustainably
- was landscaped to promote biological diversity
- promoted analytical skill in assessing full costs over the lifetime of the building
- promoted ecological competence and mindfulness of place
- became in its design and operations, genuinely pedagogical
- met rigorous requirements for full-cost accounting.

We intended, in other words, a building that did not impair human or ecological health somewhere else or at some later time.

Endorsed by a new president of the college, the project moved forward in the fall of 1995. Two graduates from the class of 1993 helped coordinate the design of the project and engaged students, faculty, and the wider community in the design process. Architect John Lyle facilitated the design charettes that began in the fall of 1995. Some 250 students, faculty, and community members eventually participated in the 13 charettes in which the goals for the center were developed and refined. From 26 architectural firms that applied for the job, we selected William McDonough & Partners in Charlottesville, Virginia.

No architect alone, however talented, could design the building that we proposed. It was necessary, therefore, to assemble a design team that would meet throughout the process. To fulfill the long-term goal that the building would eventually generate more electricity than it used, we engaged Amory Lovins and Bill Browning from the Rocky Mountain Institute as well as scientists from NASA, Lewis Space Center. To meet the standard of zero discharge, we hired John Todd and Michael Shaw, the leading figures in the field of ecological engineering. The landscape plan was developed by John Lyle and Andropogen, Inc., from Philadelphia. To this team we added structural and mechanical engineers and a contractor. During the programming and schematic design phase this team and representatives from the college met by conference call weekly and in regular working sessions.

The team approach to architectural design was a new process for Oberlin College. Typically, architects do the basic design, ask engineers to heat and cool it, and bring in landscapers to make it look pretty. By engaging the full design team from the beginning, we intended to improve the integration of building systems and technologies and the relationship between the building and its landscape. Early on, we decided that the standard for technology in the building was to be state-of-the-shelf, but within state-of-the-art design. In other words, we did not want the risk of untried technologies, but we did want the overall product to be at the frontier of what it is now possible to do with ecologically smart design.

The building program called for major changes, not only in the design process but also in the selection of materials, relationship to manufacturers, and in the way we counted the costs of the project. We intended to use materials that did not compromise human health or dignity somewhere else. We also wanted to use materials that had as little embodied fossil energy as possible, hence giving preference to those locally manufactured or grown. In the process we discovered how little is generally known about the ecological and human effects of the materials system and how little the present tax and pricing system supports standards upholding ecological or human integrity. Unsurprisingly, we also discovered that the present system of building codes does little to encourage innovation leading to greater resource efficiency and environmental quality.

Typically, buildings are a kind of snapshot of the state of technology at a given time. In this case, however, we intended for the building to remain technologically dynamic over a long period of time. In effect, we proposed that the building adapt or learn as the state of technology changed and as our understanding of design became more sophisticated. This meant that we did not necessarily want to own particular components of the building such as the photovoltaic electric system which would be rendered obsolete as the technology advanced. We explored other arrangements, including leasing materials and technologies that will change markedly over the lifetime of the building.

The same strategy applied to materials. McDonough & Partners regarded the building as a union of two different metabolisms: industrial and ecological. Materials that might eventually decompose into soil were considered parts of an ecological metabolism. Otherwise they were regarded as part of an industrial metabolism and might be leased from the manufacturer and eventually returned as a feedstock to be remanufactured into new product.

The manner in which we appraised the total cost of the project represented another departure from standard practice of design and construction. Costs are normally considered synonymous with the those of design and construction. As a consequence, institutions tend to ignore the costs that buildings incur over expected lifetimes as well as all of those other costs to environment and human health not included in the prices of energy, materials, and waste disposal. The costs of this project, accordingly, were higher than normal because we included

- students, faculty, and community members in the design process
- research into materials and technologies to meet program goals
- higher performance standards
- more sophisticated technologies

- greater efforts to integrate technologies and systems
- an endowment fund for building maintenance.

In addition, we expect to do a materials audit of the building, including an estimate of the amount of carbon dioxide released by the construction, along with a menu of possibilities to offset these costs.

The groundbreaking occurred in the fall of 1998. We occupied the building in January of 2000. We now know that the goals for the project were reasonable if ambitious. The building now generates a substantial portion of the electricity that it uses. It purifies wastewater on site. It is designed to remain technologically dynamic well into the future. It is being instrumented to report its performance data in real time on a college Web site. The landscape includes a small restored wetland and forest as well as gardens and orchards. In short, it is designed to instruct students and faculty in the arts of ecological competence and the possibilities of ecological design applied to buildings, energy systems, wastewater, landscapes, and technology, all of which are now parts of our curriculum.

As important as the building and its landscape, one of the more important effects of the project has been its impact on those who participated. Some of the students who devoted time and energy to the project began to describe it as their legacy to the college. Because of their work on the project, many of them learned about ecological design and how to solve real problems by working with some of the best practitioners in the world. Some of the faculty who participated in the effort and who were skeptical about the possibility of changing the institution came to see change as sometimes possible. And perhaps some of the college officials who initially saw this as a risky project came to regard risks incurred for the right goals as worthwhile.

Is the Adam Joseph Lewis Center a perfect building? Absolutely not. It is, however, a very good building and a beginning to much more. To paraphrase Wes Jackson (1985), relative to the potential for ecological design, this is Kitty Hawk and we're 10 feet off the ground. But someday some of the students who worked on this project will design buildings and communities that are the ecological equivalent of 747s.

The real test, however, lies ahead. It will be tempting for some, no doubt, to regard this as an interesting but isolated experiment having no relation to other buildings now in the planning stage or for campus landscaping or resource management. The pedagogically challenged will see no further possibilities for rethinking the process, substance, and goals of education. If so, the center will exist as an island on a campus that mirrors the larger culture. On the other hand, the project offers a model that might inform architectural standards for all new construction and renovation; decisions about landscape management; financial decisions about payback times and full-cost accounting; courses and projects around the solution to real problems; and how we engage the wider community.

By some estimates, humankind is preparing to build more in the next half century than it has built throughout all of recorded history. If we do this inefficiently and carelessly, we will cast a long ecological shadow on the human future. If we fail to pay the full environmental costs of development, the resulting ecological and human damage will be irreparable. To the extent that we do not aim for efficiency and the use of renewable energy sources, the energy and maintenance costs will unnecessarily divert capital from other, far better purposes. The dream of sustainability, however defined, would then prove to be only a fantasy. Ideas and ideals need to be rendered into models and examples that make them visible, comprehensible, and compelling. Who will do this?

More than any other institution in modern society, colleges and universities have a moral stake in the health, beauty, and integrity of the world our students will inherit. We have an obligation to provide our students with tangible models that calibrate our values and capabilities—models that they can see, touch, and experience. We have an obligation to create grounds for hope in our students who sometimes define themselves as "Gen X." But hope is different from wishful thinking so we have a corollary obligation to equip our students with the analytical skills and practical competence necessary to act on high expectations. When the pedagogical abstractions, words, and whole courses do not fit the way the buildings and landscape constituting the academic campus in fact work, students learn that hope is just wishful thinking or, worse, rank hypocrisy. In short, we have an obligation to equip our students to do the hard work ahead of

- learning to power civilization by current sunlight
- reducing the amount of materials, water, and land use per capita

- growing food and fiber sustainably
- disinventing the concept of waste
- preserving biological diversity
- restoring ecologies ruined in the past century
- rethinking the political basis of modern society
- developing economies that can be sustained within the limits of nature
- distributing wealth fairly within and between generations.

No generation ever faced a more daunting agenda. But none ever faced more exciting possibilities either. Do we now have or could we acquire the know-how to power civilization by sunlight or to reduce the size of the human footprint (Wackernagel and Rees 1996) or grow our food sustainably or prevent pollution or preserve biological diversity or restore degraded ecologies? In each case I believe that the answer is yes. Whether we possess the will and moral energy to do so while rethinking political and economic systems and the distribution of wealth within and between generations remains to be seen.

Finally, the potential for ecologically smarter design in all of its manifestations in architecture, landscape design, community design, the management of agricultural and forest lands, manufacturing, and technology does not amount to a fix for all that ails us. Reducing the amount of damage we do to the world per capita will only buy us a few decades, perhaps a century if we are lucky. If we squander that reprieve, we will have succeeded only in delaying the eventual collision between unfettered human desires and the limits of the earth. The default setting of our civilization needs to be reset to ensure that we build a sustainable world that is also spiritually sustaining. This is not a battle between left and right or haves and have-nots as it is often described. At a deeper level the issue has to do with art and beauty. In the largest sense, what we must do to ensure human tenure on the earth is to cultivate a new standard that defines beauty as that which causes no ugliness somewhere else or at some later time.

15

The Architecture of Science

When you build a thing you cannot merely build that thing in isolation, but must also repair the world around it, and within it, so that the larger world at that one place becomes more coherent, and more whole.

—*Christopher Alexander*

Back to the future. Suppose for a moment that you are the chair of a faculty team at Cornell University in the year 1905 and are charged with the responsibility for developing plans for a new science building. You, however, have the foreknowledge that this building is the one in which a young man from Columbus, Ohio, Thomas Midgley Jr., will one day learn his basic science. Further, you know what he will do over the course of his career. You have only this one chance to affect the mind of the man who will otherwise someday hold the world's record for banned toxic substances by formulating leaded gasoline and chlorofluorocarbons. What would you do? Before developing the building program, could you engage your faculty colleagues in a conversation about the kind of science to be taught in the building? Would it be possible, in other words, to make architecture a derivative of curriculum? Would it be possible to signal to all entering the building that knowledge is always incomplete and that, at some scale and under some conditions, it can be dangerous? Is it possible to make this warning similar to but more effective than the Surgeon General's warning on a pack of cigarettes? If you succeed, the catastrophes of lead dispersal from automobile exhaust and the thinning of stratospheric ozone from chlorofluorocarbons will not occur.

Of course, the design of science buildings alone is not likely to influence young minds as much as teachers, peers, and classes do, but it is far from inconsequential. Frank Lloyd Wright once said that he could design a house for a newly married couple that would cause them to divorce within a matter of weeks. By the same logic, it is possible to design science buildings in such a way that they contribute to the estrangement of mind and nature, deadening senses and sensibilities. Indeed, this is the way we typically construct buildings. Typically, science buildings are massive and fortresslike and give no hint of intimacy with nature. Their design is utilitarian, with long, straight corridors and graceless, square rooms. Neither daylight nor natural sounds are permitted. Windows do not open. Air, expensively heated and cooled by the combustion of fossil fuels, is forced noisily through the structure. Toxic compounds vented from laboratories drift toward neighborhoods downwind. Neither the building nor classes taught in it give any reason to question human domination of nature. Both celebrate the advance of human knowledge, giving no hint of the things we do not or cannot know and little cause for humility in the face of mystery. Accordingly, the building conveys the mistaken impression that every advance of knowledge is a defeat for ignorance. It is dedicated to one particular discipline and, if profitable, to the commercial exploitation of knowledge. Architecture in such buildings does nothing to soften or improve human relationships in such buildings that tend to reflect fear—of making a mistake, of failure to receive tenure or promotion, or merely that of anonymity. Conversation in offices, lecture halls, and corridors occurs within a narrow envelope of disciplinary language and assumptions, and often has little in common with that of the humanities. Visitors coming into such buildings often feel that they are in an alien place. On some campuses, entrance is granted only to those with a security clearance. The surrounding land-scape is paved over for parking. And it is widely believed that this is a good place for the young to learn science.

I believe that it is possible to design science buildings so well that they can help promote conventional smartness, as well as a wide-angle view of the world and a love for the creation. Architectural de-sign is unavoidably a kind of crystallized pedagogy that instructs in powerful but subtle ways. It teaches participation or exclusion. It di-rects what we see, how we move, and our sense of time and space. It affects how and how well we relate to each other and how carefully we relate to the natural systems from which we extract energy and materials and to which we consign our wastes. Most important, it in-fluences how we think and how we think about thinking. For archi-tecture to instruct in positive ways, we must be willing to question old assumptions about the human role in nature that are often embedded in the design of science buildings just as they are embedded in a cur-riculum with roots going back to Bacon, Descartes, and Galileo.

But no such assessment can take place within the safe and com-fortable confines of any single discipline. It is as much a conversation about ethics, politics, economics, and sociology that affects how knowledge is used in the world as it is about biology, chemistry, geol-ogy, or physics. It could not be conducted in the jargon of any one dis-cipline but only in the common language. It would require a high level of honesty. It is a conversation about what, given our present cir-cumstances, is worth knowing and what's not. It is, in other words, about our priorities in an increasingly perilous time in human history. Such a conversation would take time and patience, and its outcome would likely offend those inclined to defend science at all costs on the one hand and those who would abolish it on the other.

To illustrate the problem, our children now have several hundred chlorinated chemicals in their fatty tissues that do not belong there and with unknown effects (Thornton 2000). We do know, however, that cancer, reproductive problems, and behavioral disorders are in-creasing everywhere. Exposure to chemicals is ubiquitous, coming from plastics, farm chemicals, gasoline additives, carpets, building ma-terials, and lawn chemicals. Some 100,000 chemicals are in use worldwide, some of which are long-lived and can be found in routine samples of soil, air, and water. This contamination happened in large measure because of a kind of promiscuous chemistry promulgated by petrochemical companies aided and abetted by academic scientists who trained the chemists hired by petrochemical companies, and thereby influenced the larger moral, political, and social framework in which chemistry would be practiced. Many academic scientists made their peace too easily with those who used scientific knowledge carelessly. This is by no means an argument against the study of chemistry. But it does raise serious questions about the kind of chemistry we teach and the larger ecological, intellectual, moral, and political framework in which chemistry is taught and practiced. It is possible, in other words, to practice chemistry as if evolution, ecology, and ethics do not matter, but it is impossible for them not to matter.

Some will respond by saying that the chemistry we now practice, Superfund sites and all, is the best of all possible chemistries and that all of the disadvantages are merely the price we must pay for a high standard of living and the unavoidable result of advancing human knowledge. But as we learn more about the effects of exposure to chemicals as well as alternatives to chemical use, both responses ring hollow. Are there problems for which the use of chemicals is not an appropriate solution? Farming, for example, has become heavily dependent on chemicals with ominous economic, ecological, and human results. But we know of alternative and better farming methods that rely on ecological relationships, cultural information, and a sophisticated knowledge of chemistry, not petrochemicals. Is there another kind of chemistry to be taught and practiced? Some think so and believe that the model is found in the various ways that nature does chemistry. We make long-lived toxic compounds in large quantities and broadcast them by air and water. Organisms in nature, in contrast, often make toxic compounds, but in small amounts that are contained and biodegradable. In billions of years of evolution lots of strategies were tried, many of which were discarded. What remains is a set of exquisite, time-tested strategies. By comparison, industrial chemistry, about a century old, is clumsy and destructive. Accordingly, the rule of thumb ought to be that if nature did not make it, we should not either. Exceptions to that rule ought to be made cautiously, on a small scale, and for reasons that will appear to be good and sufficient to those who will eventually bear the consequences.

The standard for chemistry modeled along the lines of natural systems is no longer whether it is possible or profitable to make, but does it fit within the larger evolving fabric of life on earth. Is it toxic?

Does it break down? Do we know what it will do in the world over the long term? And where does it fit in a just, caring, and competent society? The standard would no longer simply be that of the successful experiment, but that of ecological health. A chemistry curriculum, accordingly, would feature the study of evolution, ecology, biology, politics, and ethics. It would equip students with guidelines for what elements should not be joined together or taken apart and why. Students would be required to master Marlowe's *Dr. Faustus*, Mary Shelley's *Frankenstein*, and Melville's *Moby-Dick*. Indeed, a better kind of chemistry is beginning to emerge in fields of industrial ecology and among companies pioneering concepts such as "products of service" that are returned to the manufacturer to be remade into new carpet (Benyus 1997, McDonough and Braungart 1998). But these concepts have yet to take hold in the teaching of academic chemistry or in the petrochemical industry (Collins 2001).

Lest I appear to single out chemistry unfairly, let me hasten to add that similar observations could be made of the other sciences and social sciences that too easily accommodated themselves to the defense establishment, oil companies, biotech companies, and global corporations. My point is not to establish guilt, but to propose a more scientific (which is to say, skeptical) science better suited to the task of protecting life.

We survived a century of dioxin, DDT, chlorinated hydrocarbons, Superfund sites, ozone holes, and nuclear bombs, but with a far smaller margin for error than we might have hoped for. We are entering a new era in science in which genetic engineering and biotechnology are taking center stage. Will this era prove to be less destructive? I doubt it. On the contrary, I think it has the potential to be even worse. We are on a course to repeat many of the same kinds of mistakes in biology that were made in the development of chemistry and for some of the same reasons having to do with hubris, ignorance, greed, and the reductionism that removes problems from their larger context. One can easily imagine books that will be written 50 years hence that will echo themes found in Rachel Carson's *Silent Spring* (1962), Lewis Mumford's *The Pentagon of Power* (1970), and David Ehrenfeld's *The Arrogance of Humanism* (1978).

In this light, how might the design of science facilities help us to avoid repeating old mistakes? First, the design process should begin not by addressing spatial needs and disciplinary priorities, but by rethinking the curriculum taught in the building. The overwhelming fact of our time is that we are in serious jeopardy of "irretrievably mutilating" the earth and causing "vast human misery" in the process (Union of Concerned Scientists 1992). Our students will need, in Richard Levins's words, a science that emphasizes "wholeness and process in complexly connected networks of causes that cross the boundaries of disciplines" (1998, 7). They will need the intellectual agility to combine reductionist science with a larger view of causality that includes other species, mind with body, complex interactions, and the intricate ways in which social patterns and hierarchies affect outcomes.

Because conversation at this depth is unlikely to happen in competition with classes, e-mail, fax machines, telephones, and committee meetings, the process of design must begin with faculty, students, and others meeting away from the busyness of the campus. Given the normal state of campus politics, it would be wise to engage the services of an adept facilitator. The goal is to honestly discuss the relationship between the concepts and skills that students will need to master in the coming century in order to protect and enhance life. Discussion about program details and architecture should follow. What at first appears to be a difficult and perhaps threatening conversation has the potential to generate intellectual excitement, greater collegiality, and a higher level of science education and research.

The actual building design should say to our students what we would like them someday to say to the world. Since it is irresponsible as well as foolish to waste energy, the building ought to use energy with the highest possible efficiency. Since we are nearing the end of the fossil fuel age, the building should be powered largely by advanced solar technologies. Since it is irresponsible to discharge toxic wastes, laboratories should be designed with a zero discharge standard. Since it is irresponsible to destroy forests, all wood used in the building ought to be harvested from those that are managed for long-term sustainability. Since it is irresponsible to use materials that are hazardous to manufacture, install, or discard, the building should be constructed from those that will be one day be returned to manufacturers for recycling or will decompose to make good soil. Since it is irresponsible to destroy biological diversity, the surrounding landscape should be designed to promote biological diversity. And since it is irresponsible to foster hypocrisy, the building should be designed to make the curriculum hidden in architecture and operations part of the formal curriculum. To that end, data on building energy performance, energy production, water quality entering and leaving the building, indoor air quality, and emissions should be collected and publicly displayed.

Instead of the serial design process described in chapter 14, ecological design requires bringing the architects, engineers, landscape designers, ecological engineers, energy analysts, and others together at the beginning of the project. The increased costs of front loading can be more than offset by better integration of technical systems, improved performance, and a better fit between the building and the landscape (Rocky Mountain Institute 1998). The results are greater efficiency and lower energy costs over the life of the structure. It is not enough to change the process, however, without changing the financial incentives that drive it. Fees for architects and engineers are typically calculated as a percentage of the total project costs of HVAC equipment installed in the building. There is, accordingly, little incentive to minimize project costs or to maximize efficiency. In contrast, fees can be calculated on the actual building performance so that the savings from higher levels of efficiency are shared between the institution and the designers (E Source 1992).

Finally, science buildings are almost always utilitarian, designed to be, as French architect Le Corbusier (1887–1965) would have had it, machinelike. It is essential to add another dimension to the architecture of science buildings. How, for example, might the present-day counterparts of Thomas Midgley Jr. be warned about the fallibility of human intelligence and the consequences of using knowledge carelessly? We sometimes memorialize tragedies after the fact in monuments to victims of human folly like the Vietnam Wall and the Holocaust Memorial. Art, sculpture, inscriptions, and visual displays should be used to warn students of future ecological tragedies. They should say unequivocally to eager and impressionable minds that the truth they seek is always elusive, partial, complex, and ironic; the world is not a machine and cannot be dismantled with impunity; and that whatever is taken apart for analytical convenience must be made whole again. Both architecture and curriculum should alert the young to the possibilities and limits of knowledge as well as the obligation to see that knowledge is used to good ends. Finally, the architecture of science buildings and the curriculum taught in them ought to reflect awareness of the fact that we, scientists and lay persons alike, stand at the edge of a vast mystery that exceeds human intelligence. D. H. Lawrence (Bates et al. 1993, 3) said it this way: "Water is H_2O, hydrogen two parts, oxygen one. But there is also a third thing that makes it water and nobody knows what that is." The world would be a better place had Thomas Midgley Jr. graduated knowing that neither intellectual brilliance nor technological cleverness could ever solve the riddle of the third thing.

16

2020: A Proposal

> We all live by robbing Asiatic coolies, and those of us who are
> "enlightened" all maintain that those coolies ought to be set
> free; but our standard of living, and hence our "enlightenment"
> demands that the robbery shall continue.
>
> —*George Orwell*

By a large margin 1998 was the warmest year ever recorded. The previous year was the second warmest (IPCC 2001). A growing volume of scientific evidence indicates that, given present trends, the combustion of fossil fuels, deforestation, and poor land-use practices will cause a major, and perhaps self-reinforcing, shift in global climate (Houghton 1997). With climatic change will come severe weather extremes, superstorms, droughts, killer heat waves, rising sea levels, spreading disease, accelerating rates of species loss, and collateral political, economic, and social effects that we cannot imagine. We are conducting, as Roger Revelle (quoted in Somerville 1996, 35) once noted, a one-time experiment on the earth that cannot be reversed and should not be run.

The debate about climatic change has, to date, been mostly about scientific facts and economics, which is to say a quarrel about unknowns and numbers. On one side are those, greatly appreciated by some in the fossil fuel industry, who argue that we do not yet know enough to act and that acting prematurely would be prohibitively expensive (Gelbspan 1998). On the other side are those who argue that we do know enough to act and that further procrastination will make subsequent action both more difficult and less efficacious. In the United States, which happens to be the largest emitter of greenhouse gases, the issue is not likely to be discussed in any constructive manner. And the U.S. Congress, caught in a miasma of ideology and partisanship, is in deep denial, unable to act on the Kyoto agreement that called for a 7 percent reduction of 1990 carbon dioxide levels by 2012. Even that level of reduction, however, would not be enough to stabilize climate.

To see our situation more clearly we need a perspective that transcends the minutiae of science, economics, and current politics. Because the effects, whatever they may be, will fall most heavily on future generations, understanding their likely perspective on our present decisions would be useful to us now. How are future generations likely to regard various positions in the debate about climatic change? Will they applaud the precision of our economic calculations that discounted their prospects to the vanishing point? Will they think us prudent for delaying action until the last-minute scientific doubts were quenched? Will they admire our heroic devotion to inefficient cars and sport utility vehicles, urban sprawl, and consumption? Hardly. They are more likely, I think, to judge us much as we now judge the parties in the debate on slavery prior to the Civil War.

Stripped to its essentials, defenders of the idea that humans can hold other humans in bondage developed four lines of argument. First, citing Greek and Roman civilization, some justified slavery by arguing that the advance of human culture and freedom had always depended on slavery. "It was an inevitable law of society," according to John C. Calhoun, "that one portion of the community depended upon the labor of another portion over which it must unavoidably exercise control" (W. L. Miller 1998, 132). And "Freedom," the editor of the *Richmond Inquirer* once declared, "is not possible without slavery"

(Oakes 1998, 141). This line of thought, discordant when appraised against other self-evident doctrines that "all men are created equal," is a tribute to the capacity of the human mind to simultaneously accommodate antithetical principles. Nonetheless, it was used by some of the most ardent defenders of "freedom" up to the Civil War.

A second line of argument was that slaves were really better off living here in servitude than they would have been in Africa. Slaves, according to Calhoun "had never existed in so comfortable, so respectable, or so civilized a condition as that which it now enjoyed in the Southern States" (W. L. Miller 1998, 132). The "happy slave" argument fared badly with the brute facts of slavery that became vivid for the American public only when dramatized by Harriet Beecher Stowe in *Uncle Tom's Cabin*, published in 1852.

A third argument for slavery was cast in cost-benefit terms. The South, it was said, could not afford to free its slaves without causing widespread economic and financial ruin. This argument put none too fine a point on the issue; slavery was simply a matter of economic survival for the ruling race.

A fourth argument, developed most forcefully by Calhoun, held that slavery, whatever its liabilities, was up to the states, and the Federal government had no right to interfere with it because the Constitution was a compact between independent political units. Beneath all such arguments, of course, lay bedrock contempt for human equality, dignity, and freedom. Most of us, in a more enlightened age, find such views repugnant.

While the parallels are not exact between arguments for slavery and those used to justify inaction in the face of prospective climatic change, they are, perhaps, sufficiently close to be instructive. First, those saying that we do not know enough yet to limit our emission of greenhouse gases argue that human civilization, by which they mean mostly economic growth for the already wealthy, depends on the consumption of fossil fuels. We, in other words, must take substantial risks with our children's future for a purportedly higher cause: the material progress of civilization now dependent on the combustion of fossil fuels. Doing so, it is argued, will add to the stock of human wealth that will enable subsequent generations to better cope with the messes that we will leave behind.

Second, proponents of procrastination now frequently admit the possibility of climatic change, but argue that it will lead to a better world. Carbon enrichment of the atmosphere will speed plant growth, enabling agriculture to flourish, increasing yields, lowering food prices, and so forth. Further, while some parts of the world may suffer, a warmer world will, on balance, be a nicer and more productive place for succeeding generations.

Third, some, arguing from a cost-benefit perspective, assert that energy conservation and solar energy are simply too expensive now. We must wait for technological breakthroughs to reduce the cost of energy efficiency and a solar-powered world. Meanwhile we continue to expand our dependence on fossil fuels, thereby making any subsequent transition still more difficult.

Finally, arguments for procrastination are grounded in a modern-day version of states' rights and extreme libertarianism which makes squandering fossil fuels a matter of individual rights, devil take the hindmost.

Of course, we do not intend to enslave subsequent generations, but we will leave them in bondage to degraded climatic and ecological conditions that we have created. Further, they will know that we failed to act on their behalf with alacrity even after it became clear that our failure to use energy efficiently and develop alternative sources of energy would severely damage their prospects. In fact, I am inclined to think that our dereliction will be judged a more egregious moral lapse than that which we now attribute to slave owners. For reasons that one day will be regarded as no more substantial than those supporting slavery, we knowingly bequeathed the risks of global destabilization to all subsequent generations everywhere. If not checked soon, that legacy will include severe droughts, heat waves, famine, changing disease patterns, rising sea levels, and political and economic instability. It will also mean degraded political, economic, and social institutions burdened by bitter conflicts over declining supplies of fossil fuels, water, and food. It is not far-fetched to think that human institutions, including democratic governments, will break under such conditions.

Other similarities exist. Both the use of humans as slaves and the use of fossil fuels allow those in control to command more work than would otherwise be possible. We no longer use slaves but we do have, on average, the fossil fuel equivalent of 75 slaves at our service (Mc-Neill 2000, 16). Both practices inflate wealth of some by robbing others. Both systems work only so long as something is underpriced: the

TABLE 16.1. A Comparison of Slavery and Procrastination on Efforts to Limit Greenhouse Gas Emissions

Issue	Argument for slavery	Argument for procrastination
Progress	Historically necessary for human improvement	Energy consumption necessary for economic growth
Improvement	Slaves better off here	A carbon-enriched world will be better for agriculture
Cost-benefit	The southern economy depends on slavery	Costs of energy efficiency are too great to bear; let's wait for better technology
Rights	The federal government's rights stop at states' borders	The rights of present-generation carbon emitters trump those of all others

devalued lives and labor of a slave or fossil fuels priced below their replacement costs. Both require that some costs be ignored: those to human beings stripped of choice, dignity, and freedom or the cost of environmental externalities, which cast a long shadow on the prospects of our descendants. In the case of slavery, the effects were egregious, brutal, and immediate. But massive use of fossil fuels simply defers the costs, different but no less burdensome, onto our descendants, who will suffer the consequences with no prospect of manumission. Slavery warped the politics and cultural evolution of the South. But our dependence on fossil fuels has substantially warped and corrupted our politics and culture as well. Slaves could be manumitted; victims of global warming have no such prospect. We leave behind steadily worsening conditions that cannot be altered in any time span meaningful to humans.

Both slavery and fossil fuel–powered industrial societies require a mass denial of responsibility. Slave owners were caught in a moral quandary. Their predicament, in James Oakes's words, was "the product of a deeply rooted psychological ambivalence that impels the individual to behave in ways that violate fundamental norms even as they fulfill basic desires" (1998, 120). Regarding slavery, George Washington confessed that "I shall frankly declare to you that I do not like even to think, much less talk, of it" (ibid., 120). As one Louisiana slave owner put it, "A gloomy cloud is hanging over our whole land" (ibid., 110). Many wished for some way out of a profoundly troubling reality. Instead of finding a decent way out, however, the South created a culture of denial around the institutions of bondage. Southerners were enslaved by their own system until it came crashing down around them in the Civil War.

We, too, find ourselves in a quandary. From poll data we know that most Americans believe that global warming is real and that its consequences could be tragic and irreversible. But the response of Congress and the business community has been to deny that the problem exists and continue with business as usual. Proposals for higher gasoline taxes, increasing fuel efficiency, or limits on use of automobiles, for example, are regarded as politically impossible as the abolition of slavery was in the 1830s. Unless we take appropriate steps soon, our system, too, will end badly.

We now know that heated arguments made for the enslavement of human beings were both morally wrong and self-defeating. The more alert knew this early on. Benjamin Franklin noted that slaves "pejorate the families that use them; the white children become proud, disgusted with labor, and being educated in idleness, are rendered unfit to get a living by industry" (Finley 1980, 100). Thomas Jefferson knew all too well that slavery degraded slaves and slave owners alike, while providing no sustainable basis for prosperity in an emerging capitalist economy. On one hand, it is possible that the extravagant use of fossil fuels has become a substitute for intelligence, exertion, design skill, and foresight. On the other hand, we have every reason to believe that vastly improved energy efficiency and an expeditious transition to a solar-powered society would be to our advantage, morally and economically. Energy efficiency could lower our energy bill in the United States alone by as much as $200 billion per year (Hawken et al. 1999). It would reduce environmental impacts associated with mining, processing, transportation, and combustion of fossil fuels and promote better technology. Elimination of subsidies for fossil fuels, nuclear power, and automobiles would save tens of billions of dollars each year (Myers 1998). In other words, the "no regrets" steps necessary to avert the possibility of severe climatic change, taken for sound ethical reasons, are the same steps we ought to take for reasons of economic self-interest. History rarely offers such a clear convergence of ethics and self-interest.

If we are to take this opportunity, however, we must be clear that the issue of climatic change is not, first and foremost, a matter of economics, technology, or science, but rather a matter of principle that is best seen from the vantage point of our descendants. The same historical period that gave us slavery also gave us the principles necessary to abolish it. What Thomas Jefferson called "remote tyranny" was not merely tyranny remote in space, but in time as well—what has been termed "intergenerational remote tyranny." In a letter to James Madison written in 1789 (Jefferson 1975, 444–451), Jefferson argued that no generation had the right to impose debt on its descendants, for were it to do so the future would be ruled by the dead, not the living.

A similar principle applies in this instance. Drawing from Jefferson, Aldo Leopold, and others, such a principle might be stated thus:

> No person, institution, or nation has the right to participate in activities that contribute to large-scale, irreversible changes of the earth's biogeochemical cycles or undermine the integrity, stability, and beauty of the earth's ecologies, the consequences of which would fall on succeeding generations as a form of irrevocable remote tyranny.

Such a principle will likely fall on uncomprehending ears in Congress and in most corporate boardrooms. Who, then, will act on it? Who ought to act? Who can lead? What institutions represent the interests of our children and succeeding generations on whom the cost of present inaction will fall? At the top of my list are those that educate and thereby equip the young for useful and decent lives. Education is done in many ways, the most powerful of which is by example. The example the present generation needs most from those who propose to prepare them for responsible adulthood is a clear signal that their teachers and mentors are responsible and will not, for any reason, encumber their future with risk or debt—ecological or economic. And they need to know that our commitment is more than just talk. This principle can be stated in these words:

> The institutions that purport to induct the young into responsible adulthood ought themselves to operate responsibly,

which is to say that they should not act in ways that might plausibly undermine the world their students will inherit.

Accordingly, I propose that every school, college, and university stand up and be counted on the issue of climatic change by beginning now to develop plans to reduce and eventually eliminate or offset the emission of heat-trapping gases by the year 2020.

Opposition to such a proposal will, predictably, follow along three lines. The first line of objection will arise from those who argue that we do not yet know enough to act. In other words, until the threat of climatic change is clear beyond any possible doubt (and also less easily reversed), we cannot act. Presumably, these same people do not wait until they smell smoke in the house at 2 A.M. to purchase fire insurance. A "no regrets" strategy relative to the far-from-remote possibility of climatic change is, by the same logic, a way to insure our descendants against the possibility of disaster otherwise caused by our carelessness.

A second line of objection will come from those who will argue that educational institutions on their own cannot afford to act. To be certain, there will be initial expenses, but there are also quick savings from reducing energy use. In fact, done smartly, implementation of energy efficiency and solar technology can save money. Moreover, it is now possible to use energy service companies that will finance the work and pay themselves from the stream of savings, making the transition budget neutral. The real problem here has less to do with costs than with moral energy and the failure to imagine possibilities in places where imagination and creativity are reportedly much valued.

A third kind of objection will come from those who agree with the overall goal of stabilizing climate, but will argue that our business is education, not social change. This position is premised on the quaint belief that what occurs in educational institutions must be uncontaminated by contact with the affairs of the world and that we have no business objecting to how that world does its business. It is further assumed that education occurs only in classrooms and must be remote from anything having practical consequences. Were the effort to eliminate the use of fossil fuels, however, done as a 20-year effort in which students worked with faculty, staff, administration, energy engineers, and technical experts, the educational and institutional benefits would be substantial. How might the abolition of fossil fuels occur? In outline, the steps are straightforward, requiring (1) a thorough audit of current institutional energy use; (2) preparation of a detailed engineering plans to upgrade energy efficiency and eliminate waste; (3) development of plans to harness renewable energy sources sufficient to meet campus energy needs by 2020; and (4) competent implementation. These steps ought to engage students, faculty, administration, staff, and representatives of the surrounding community. They ought to be taken publicly as a way to educate a broad constituency about the consequences of our present course and the possibilities and opportunities for change.

Some colleges are beginning to act on climate change. Fifty-six college presidents in New Jersey agreed to meet or exceed the Kyoto Protocol. Tufts University has launched a "Cool Planet, Clean Air" initiative with an alliance of New England colleges and universities. Oberlin College, working with the Rocky Mountain Institute, has completed a study of what would be required for the institution to become "climatically neutral" by the year 2020. The longer-term goal of such efforts is to begin, from the grass roots, the long-delayed transition to energy efficiency and solar power. Perhaps our leaders will follow one day when they are wise enough to distinguish the public interest from narrow, short-run private interests. Someday, too, all of us will come to understand that true prosperity neither permits nor requires bondage of any human being, in any form, for any reason, now or ever.

17

Education, Careers, and Callings

In the past decade I have received several hundred letters of inquiry from students asking for advice about education and careers. Most want to know how to combine their passion for the natural world with formal education in order to craft a useful life. The letters and e-mails are often written in a tone of frustration. An undergraduate biology major, for example, writes: "I have been researching my options, and I have come to the conclusions [*sic*] that there are quite a number of programs labeled 'conservation biology' or 'environmental studies' around the country. It is fairly easy to become lost in a sea of them. I attended the Society for Conservation Biology meeting in Maryland, but failed to find any prospective advisors. Would you have any advice to offer on this topic?" Similarly, a recent Ph.D. in wildlife biology writes: "I am struggling to translate my professional training into a life well lived that in some way might contribute to preserving the natural world and not just documenting its decline. . . . My professional training did not prepare me well for these tasks." Dozens of other letters have the same plaintive themes.

The problem is not simply that there are many more students who want practical careers in environmental work than those who find them. The deeper problem has to do with the experience of students as they pass through the system of higher education. Whatever they once may have been, institutions of higher education have become vast and expensively operated machines much like any for-profit corporation. Students are fed through a conveyor belt of requirements, large classes, deadlines, and general busy-ness. What they learn seldom adds up to anything like a coherent, ecologically solvent worldview. The scale of most institutions is not conducive to humane interaction. Seldom encouraged to discern an inner calling, students are more often counseled to find secure careers that pay well. Nonetheless, many students still feel a calling toward service that runs counter to the incentives, values, and structure of their formal education.

This was brought home to me during a recent conference to review various fellowship programs, including some in conservation biology, offered through prominent universities. Without question, fellowships such as these have helped a number of young scholars complete their graduate work and move into professional careers. Judged by most conventional criteria, all the programs we reviewed have been successful. The proceedings, however, were permeated by a sense of self-congratulation that seemed oddly remote from the larger backdrop of global trends. When asked, for example, how she defined success, one participant replied that success meant "well-trained students who finish their Ph.D.s on time." Another argued that "depth and rigor in a particular field promoted collegial interaction across disciplines," a view that would astonish many in higher education. While a third agreed that it had taken his field a long time to discover a connection with the environment, that tardiness required no further explanation or analysis. A fourth noted that graduate studies seldom generated a "critical class" of scholars, but found that unworthy of further comment. Over and over again, the word "training" (what one does to a dog) was used where the appropriate word would have been "education." This is a great deal more than a semantic quibble. It represents a view of learning and higher education that deserves to be challenged. The university participants, good people all, regard themselves as "professionals," perhaps even as knowledge technicians. Under the right circumstances this pays well and provides indoor employment, but it can also generate bullet-proof complacency.

In a subsequent analysis, one conference participant voiced the opinion that the problems of the world will be solved only by "detailed knowledge . . . created through empirical research" disseminated by universities. Accordingly, we provide "young people in the first stages of careers . . . a perspective for understanding complex systems and a basis for developing analytic skills." Similarly, it was assumed that "academic institutions . . . confer prestige and legitimacy" otherwise not available and that this is necessary for those embarking on professional careers. And the fact that conventional, discipline-based university programs were often "stultifying" was thought to be a minor problem.

On reflection, I think that it is a mistake to presume that what ails the world has much to do with a lack of empirical knowledge, a shortage of information, or a scarcity of professional, career-oriented scholars. It is likely that we suffer far more from a lack of courage, good-heartedness, creativity, and a larger vision of how we might integrate human societies into natural systems. But these traits are not often rewarded or even recognized in places that dispense prestige and legitimacy. On the contrary, such traits are often penalized in such places. In large part the reasons are to be found in the close relationship between the modern university and particular disciplines with corporations promoting, among other things, agribusiness, genetic engineering, artificial intelligence, the consumer economy, weapons research, and the excessive resource extraction necessary to all of the above.

One of our charges was to consider the adequacy of financial support for various fellowship programs. But funding in institutions with billion-dollar endowments is seldom a problem . . . for the things that are valued in such places. The problem is that many essentials of the long-term health of the world in which our students will live are seldom high on the priority list of institutions of higher education. As a result, many facets of a long-term perspective go begging, while parking decks, athletic facilities, and administrators flourish like mushrooms after a spring rain. What often appears as a funding problem is first and foremost a problem of values and priorities, and alert students are aware of the difference.

A few of us hoped to find programs that encouraged fellowship recipients to boldly cross the boundaries of disciplines and to connect fields of knowledge. Alas, the modern university facilitates interdisciplinary work with about the same gusto and creativity as some Balkan countries facilitate interregional tourism. The reasons are many. As a result, we often launch promising young people into academic careers that eviscerate their idealism and energy. By the time students have enrolled in graduate programs, they will have made major decisions about their career path. Soon thereafter they will have been socialized into the ethos of graduate school by a combination of fear of failure, financial dependency, and the asymmetrical power relationships that pervade such places. To succeed, they must invest more than just money and time in an effort to get a Ph.D. They must buy into a particular worldview congenial to professionalized, disciplinary knowledge and institutionalized science. Dissidents are mostly invited out. By the time a student has been exposed to four years of college and four or more years of graduate school, the psychological investment is large, as is the investment in time and money. There should be no great mystery why such systems do not turn out a higher percentage enrolled in the "critical class" of scholars who are able and willing to critique the kinds of knowledge generated in some of our proudest institutions of higher learning and how such knowledge is used.

There is a related problem. Most of us hope that environmental science will provide more than a rigorous documentation of biotic impoverishment. If so, we must be open to the disconcerting possibility that the lens of Western science distorts as often as it clarifies. Describing the ways by which the native Yup'ik people of Alaska understand nature, for example, historian Calvin Martin writes, "Their call for respect for old ways has no soil, in our reality, to take root and grow" (1999, 111). To minds that perceive reality as participatory, Western-style research is "strange, discourteous, and vaguely dangerous. . . . There is a crazy objectification going on here" by which animals "are removed from the individual's experience with them [and rendered] into 'resources' or 'objects' to be 'managed' or 'studied'" (ibid., 112). The heart of the issue for Martin lies in the choice we make between measuring the world in fear or in trust. "That decision appears to usher its bearer inexorably into one realm of realty or another, mutually exclusive of one another" (ibid., 205). Such observations bring us to an inconvenient truth that other cultures armed with far less hard science but much more of what we disparage as myth have made far better management decisions than we have.

If there is some fatal flaw in a science intent on "enlarging the bounds of the human empire, to the effecting of all things possible" as Francis Bacon put it in *New Atlantis* (1627, 447), how would a student in a typical college or university come to recognize it? How would they learn to see the dangers in, say, efforts to reengineer the gene pool of the planet? Or those to displace humans with machines that will be vastly more "intelligent"? How would they learn the humility, compassion, and perspective that should discipline the search for knowledge and its use? Could they learn to trust the world like the Yup'ik?

All of this is a way of asking, if graduate training is the solution, what is the problem? Do we intend to perpetuate an academic system well integrated with the status quo, or do we wish to preserve the earth's biota? The relation between the values built into the machinery of higher education and the values that animate most students seeking careers in conservation is not great. What alternatives could be created?

A Modest Proposal

If we intend to turn out not just scholars, but whole people who create and use knowledge to make a difference, how would we do it? Is there a teachable moment in the lives of students who want careers in conservation biology? I believe that there is, and that it most often occurs between the undergraduate experience and graduate school. At this point in their development, most young people have a fair grasp of one or more disciplines but only a vague idea of what they want to do with their lives. Many have taken a lien on future income to pay for their undergraduate degree. At that point, however, their choices generally narrow down to staying in school supported by a combination of scholarships and loans or employment. But a small minority goes on to the Peace Corps and other service organizations, often with illuminating results. Most describe such experiences as life changing, because of exposure to different cultures, ideas, and particular persons. The impact of such exposure has little to do with formal learning and everything to do with coming to see the world through different eyes. Whether they go on to graduate school or employment, most have been profoundly deepened by the experience and understand themselves and the world in ways not otherwise possible.

This suggests a possible alternative to the standard academic career track. The time between undergraduate education and graduate school is a great and mostly untapped time to influence young people before they commit to one career or another. What do they need? More than further exposure to the professoriate, they need exposure to people doing great things with courage, stamina, and creativity. They need mentors and role models, and these are most often found among those actually changing the world. Instead of career planning, they need a deeper and more vivid concept of what it means to live a life of service and commitment in what surely will be the most fateful period in human history. They need a compass to chart a life course that combines intellect, heart, judgment, and professional skills. There are a few precedents for this kind of experience, including the Watson Fellowship program and the Ashoka Network of social entrepreneurs assembled by Bill Drayton (Bornstein 1998).

I propose that such models be used to develop programs that broker a mentoring arrangement between undergraduates wanting careers in conservation and a group of extraordinary practitioners in the field. Such a program would entail the development of a selection process to identify applicants; the selection of a group of conservation practitioners; the creation of an application process that would match the two; and administration and assessment.

Selection

Such programs ought to be made available to students wanting to pursue careers in conservation. They could be identified by a discerning group of nominators—faculty as well as people in the field working in parks, wildlife refuges, environmental not-for-profit organizations, and wilderness areas.

Mentors

I propose that mentors be nominated by people who are in a good position to know who's doing what around the world. In this category I would include foundation officers, newspaper reporters, members of nongovernmental organizations, government officials, and clergy. The goal is to identify people of significant stature and accomplishment working at the intersection of conservation and human improvement with courage, stamina, and often with little public acknowledgment.

Process

The process would entail several steps. First, nominees would submit an application describing their background, interests, and ambitions. Those selected at this first stage would be sent a list of mentors with detailed information about their work. In a second submittal applicants would describe the mentors with whom they wish to spend time, the reasons for doing so, and the nature of the final product of their journey. Any applicant could identify an itinerary that would involve some time with people on the list of mentors. A third stage in the process would entail brokering a working relationship with particular mentors and their organizations. At regular intervals during the year or two, applicants and mentors would submit reports about their progress. At the end of the designated time, recipients would submit their final project that might take various forms: a book, journal, reports, articles, or documentary film.

Administration

A program of the kind outlined above would require competent, discerning, and cost-effective administration, hardly typical of higher education. It would require a larger view of what it means to be qualified to teach than mere possession of academic credentials. It would require discerning judgment about young people and their capabilities and potentials. For these reasons, I think that any such program ought to be run by a resourceful, agile, and well-connected nonprofit organization or a consortium of such organizations.

Summary

The founders of ecology and hybrid fields like conservation biology intended these to be revolutionary enterprises that joined good science with the application of knowledge (Sears 1964; Shepard and McKinley 1969; Soule 1986). In the intervening years these fields have indeed come a long way. But as academic endeavors, these disciplines are situated within institutions that have yet to demonstrate a substantial commitment to an ecologically viable future. Ours is the age-old problem of trying to put new wine into old wineskins: we have a revolutionary credo about human responsibilities for the natural world, but we mostly work in institutions still dedicated to the task of extending human mastery over the world. I know of no one solution for this problem, but there are things that can be done to expand the ecological imagination of our students, to stretch their sense of possibilities, and to connect them to people changing the world.

Postscript: The fellowship described here has been created by the Compton Foundation, Menlo Park, California. The first class of fellowship recipients will begin their terms in Spring 2002.

18

A Higher Order of Heroism

In the towns and cities across America, it is common to find a town square with a large monument to one military hero or another. Seldom, however, does one find the designers of those towns or town squares similarly memorialized. A smarter and more durable society would first acknowledge those with the foresight and dedication to design our places well, not just those who defended them in times of trouble. We need to recognize a higher order of heroism—those who helped avoid conflict, harmonized human communities with their surroundings, preserved soil and biological diversity, and created the basis for a more permanent peace than that possible to forge by violence. These are quiet heroes and heroines who work mostly out of the light of publicity. The few who do receive public acclaim are mostly reticent about the attention they get. Some like Frederick Law Olmsted, Aldo Leopold, and Rachel Carson develop a wide international following. Most, however, labor in obscurity, content to do their work for the satisfaction of doing things well. John Lyle, professor of landscape architecture at California Polytechnic Institute, was such a man.

I met John in the mid-1980s during a visit to Cal Poly. During the two days we spent together, we talked about his concept of regenerative design and his plans for the Center for Regenerative Studies, now named the Lyle Center, and walked over the site—located between a large landfill and the university. In subsequent years, John and I met at conferences and sometimes collaborated on design projects, including one located in a remote, hilly, southern rural community. Our first site visit coincided with an ice storm the previous day that had covered the region with an inch of ice. We got within a mile of the site in a rental car, but had to make our way down a long, steep hill with a sheer drop of several hundred feet on one side. For the final mile on what passed for a dirt road in that part of the country, the rental car was useless, so we began to slip, slide, and tumble our way down the hill. Near the bottom, the road banked steeply to the right, but we had to reach a trail on the left side. There was no way to walk across that ice-covered dirt road to the other side, so we did what professionals in our circumstances are trained to do: we crawled across the ice on our hands and knees. Midway, hands bleeding, John turned to me and said, "I don't mind crawling this way, or even getting run over by a pickup truck, but I sure hope no one sees us." We both laughed so hard that we lost our grip on the ice and slid backward into the ditch. Later that day I learned that John had diabetes.

When I began the project described in chapter 14, John was the first person I called to help organize the effort behind what later became the Adam Joseph Lewis Center at Oberlin College. John's dedication to that project was legendary. Flying from California, he would usually arrive in Oberlin about midnight, but would be ready to work by 8 A.M. the next morning. On more than one occasion he arrived in town too late to get a hotel room and spent the night in a rental car or on whatever spare couch he could find, always without a whisper of complaint. John was that kind of person—modest, diligent, self-denying, creative, and supportive of those around him.

I talked with John in the spring of 1998 before I left on a trip to Greece. He had a nagging cough and was scheduled for a checkup. On my return I called to inquire how he was feeling. "They've given me two weeks to live," he replied. Stunned, I sat down to write a farewell letter to a man I'd come to depend on as a valued colleague, friend, and mentor. Words at times like that are utterly inadequate, but they're all we have. That letter read in part: "The Oberlin project simply would not have happened without your dedication and quiet competence from the very beginning. In more ways than I can recount, you held things together. You were a rock throughout the entire effort. For that and for all of the late-night trips to Oberlin, the untold hours of work on the landscape design and on the entire project—thank you, thank you, thank you." When my mind goes back to John Lyle, it is always with gratitude for the time spent with him and for the example of his life. Before he died, Oberlin College named the plaza in front of the Adam Joseph Lewis Center the John Lyle Plaza.

On the Cal Poly campus John Lyle's legacy is the Center for Regenerative Studies—the facility that he helped conceive and develop. The center represents the manifestation of his thought about architecture, integrated design, and the educational process, as well as an utterly clear-headed view of the human predicament in the twenty-first century. John's professional work, both written and built, is a legacy in the form of a challenge to the conventional wisdom of our time. Trained as an architect and landscape architect, John was a pioneer in a new and more encompassing field of ecological design that embraced virtually all of the liberal arts. He left behind a body of ideas in two remarkable books and dozens of articles. That portion of his legacy comes as a challenge to all of us, but especially to educational institutions.

First, John Lyle challenged us to face the fact that "we have created a world that is simultaneously growing out of control and progressively destroying itself" (1997, 1). A world designed around linear flows will, in due course, come to ruin. As a result, this generation of students will live in a radically altered world. Sometime in 2001 world population passed 6 billion, and it may reach 8–10 billion within the lifetime of a current university student. Given present trends of species loss, these young people will live in a steadily more biologically impoverished world. Estimates vary, but it is not inconceivable that 15–20 percent of the species now extant will disappear within the next 60 years, with consequences that we cannot know. This will be the first generation ever to experience human-driven climatic change and with it increased storms and storm damage, rising sea levels, droughts, heat waves, spreading diseases, and political turmoil. These and other trends will interact in ways we will not foresee. All of this is to say that the rising generation will live in far more volatile and stressful world than any previous generation. And none has ever faced a more daunting agenda.

But Lyle's legacy to us is not one of despair, denial, or wishful thinking built on fantasies of heroic technologies or salvation by economic growth. It is, rather, one of hope founded on more solid ground. Lyle was an optimist who believed that "what humans designed we can redesign and what humans built, we can rebuild" (Lyle 1997, 2). If we act wisely, the future would be better than that which is now in prospect (Lyle 1994, 12). To act wisely means making our actions conform to ecological realities. To that end Lyle proposed to equip people to become ecologically competent by understanding the physical processes, energy flows, landforms, and the biota of the places where they lived.

Second, Lyle challenged us to deal with the structure of what ails us, not merely the rates of change. "The problems," he wrote, "are manifestations of structural failure in the global infrastructure" (1994, 9). In our circumstances, neither half-measures nor Band-Aid solutions will do. The vast infrastructure of steel, chemicals, and concrete characteristic of the modern world would have to be replaced with, as he put it, "neotechnic" solutions that are regenerative. Regeneration implies "replacing the present linear systems of throughput flows with cyclical flows" and moving "to a [world] rooted in natural processes" (ibid., 10–11). Regenerative systems would slow the velocity of water and materials, replacing machines with landscape. In such a world "mind and nature join in partnership" (ibid., 27).

Few have thought more deeply or more practically about what such a partnership with nature would mean. Lyle's vision of regenerative design was founded on 12 principles:

- Let nature i.e. natural processes do the work for us.
- Use nature as the model for human enterprise.
- Aggregate functions and processes to create resilience.
- Strive for optimum levels, not maximum.
- Match technology to needs.
- Replace power with information.
- Provide multiple pathways.
- Solve many problems simultaneously.
- Manage storage as a key to sustainability.

- Shape form to guide flow.
- Shape form to manifest process.
- Prioritize for sustainability.

For Lyle, these were not simply abstract principles, but guidelines for the development of the Center for Regenerative Studies and other projects in which he was engaged. In a larger context, the principles of regeneration were the blueprint for a society that would be powered by sunshine and grounded in the facts of nature, not grand ideologies or abstract economic theories. Consequently, society would operate at a scale, speed, and elegance fitted to natural systems. For Lyle, a regenerative society was not austere, but richer in experience, satisfaction, and conviviality.

In *Regenerative Design for Sustainable Development* (1994), Lyle described in great detail how a better and more sustainable society could provision itself with energy, materials, food, shelter, and cycle its waste. He did not stop with technical details but went on to the harder issue of politics. The largest obstacle to sustainable development was the "concentrat[ion] of power and resources among a very small number of people" (ibid., 264). Because they are smaller in scale, dispersed, and modular, regenerative technologies do not lend themselves so easily to the concentration of power. Rather than rely on the long-distance transport of energy, water, and materials, a regenerative society would make "their life support systems . . . integral parts of the local landscape" (ibid., 266). Power and wealth in that society would be more dispersed.

How would a truly regenerative society come into existence? "How do we educate the mind in nature?" (Lyle 1994, 269). The crux of the matter is to change our manner of thinking, and this means changing both the substance and process of education to join art and science. The curriculum evolving at the Center for Regenerative Studies draws from many sources, including the work of John Dewey, but mostly Lyle thought it should emerge from the experience of the enterprise itself. Education in a "paleotechnic" society, Lyle wrote, "tends to focus on products, treating them as if they were frozen in time." But in an ecological perspective, "all that exists is in process" (ibid., 270). Education appropriate to a neotechnic society would begin with the basic facts of change and interconnectedness. But how do we change educational institutions that have, as he put it, "strong tendencies toward rejection" (ibid., 273) of new ideas and integrative purposes? Lyle posed the question, but others will have to answer it. His role was to initiate the Center for Regenerative Studies in the faith that it would become part of a larger process of educational regeneration grounded in place and aiming toward permanence.

Lyle's strategy was rendered visible in the development of the Center for Regenerative Studies at Cal Poly, which he intended this to be a working model for students, faculty, and administrators. The center was to be more than an island sealed off within a larger structure. Lyle intended, rather, to change the very DNA of the institution, altering its evolution in order to engage the deep problems of our time. The subsequent history of the effort is unsurprising except for the fact that the vision has survived despite differences over administration and purposes. These are not, I think, unusual. A worldview rooted in the principles of regeneration is unavoidably at odds with the extractive mindset of the industrial order. Similarly, a curriculum that equips students for lives in a world that is ecologically durable runs counter to one that aims to equip students for success in a failing paleotechnic society. Implicit in Lyle's work is the challenge to find common ground between these two views in order to build a world that is ecologically solvent while retaining the hard-won advantages of an open and free society.

Lyle ended *Regenerative Design* by relating the potential for regenerating larger systems, cities, regions, and entire economies. His aim was to forge the links between locality and geographic regions and between ecology and an ecologically robust economics. He recognized that processes of degeneration were rooted in pre-ecological theories of economics and in massive subsidies to extractive industries. Regenerative solutions that worked with the ecology of specific places seldom received federal subsidies or research funding. He recognized the need for a larger revolution in the conduct of national and international affairs built on a more honest accounting of the costs of what we do.

Lyle's legacy is that rarest of gifts: the example of an honest and searching mind uncluttered by trivialities or intellectual fashion. His scholarship gives testimony to his remarkable breadth of knowledge and the clarity of his mind. But John Lyle was no pedant. He aimed, rather, to harness knowledge and research to improve the human

TABLE 18.1. Conflicting Paradigms: Paleotechnic versus Neotechnic

	Paleotechnic	Neotechnic
Worldview	Industrial	Ecological
Scale	Large	Small
Scope	Narrow	Integrated
Power	Concentrated	Dispersed
Wealth	Concentrated	Dispersed
Energy	Fossil fuels	Sunlight
Planning	Fragmented	Integrated
Solutions	Technological	Ecological/community
Knowledge	Concentrated	Dispersed
Accounting	Start-up costs	Life cycle

prospect by grounding it in the ecological realities of particular places and landscapes.

Lyle gave us a model of a better kind of education. He was an educator in the best sense of the word. In my experience with him over 15 years in various projects and settings, he never imposed, but rather quietly educed, which is to say, he brought forth ideas from his students and colleagues. He had an ecological view of learning which focused on process, interaction, and, above all, the power of good example. Lyle challenged his students in the 606 design studio and all of us to make something real of our ideas and to take responsibility for how those ideas are used in the world.

Lyle helped develop a larger response to the world in what he called environmental design, which is "where the earth and its processes join with human culture and behavior to create form . . . where people and nature meet where art and science join" (1994, ix). Design, the art of making things that fit harmoniously in an ecological context, is now beginning to inform architecture, landscape architecture, urban planning, business, and economics. Lyle played a key role in what, I believe, later generations will regard as the ecological enlightenment that began in the final quarter of the twentieth century.

Finally, Lyle's legacy to us includes the example of a life lived with grace, stamina, and purpose. All of his colleagues, students, and clients would agree. Lyle combined exemplary professional skill, personal humility, kindness, and dogged determination. He joined style and substance to do the right things in the right way. The power of his work came from the synergy of steadiness and vision. He showed everyone who knew him that largeness of vision could and should come from largeness of spirit.

By all standards, John Lyle left behind a remarkable legacy. But what will institutions of higher education make of it? One answer is that it will be largely ignored in the same way that a body rejects a transplanted organ by sealing it off. The Lyle Center for Regenerative Studies would then be merely a museum of quaint ideas and technologies, but not the start of something fundamentally regenerative. On the other hand, the center could grow to be a transforming force throughout higher education. Lyle challenged us to talk and listen across the barriers of different intellectual perspectives and disciplines and to transcend the routines of hierarchical management and the pettiness that often pervades academic politics. He challenged us to develop a curriculum that joins head, hands, and heart and thereby make education an agent of regeneration in the world. But most important, John Lyle left his example of a man responding to the challenges of our time with good heart, imagination, professional skill, and hope.

§ 5

CHARITY, WILDNESS, AND CHILDREN

19

The Ecology of Giving and Consuming

What one person has, another cannot have. . . . Every atom of
substance, of whatever kind, used or consumed, is so much
human life spent.

—John Ruskin

How do we sell more stuff to more people in more places?

—IBM advertisement

Don't try to eat more than you can lift.

—Miss Piggy

Some years ago a friend of mine, Stuart Mace, gave me a letter opener
hand-carved from a piece of rosewood. Over his 70-some years Stuart
had become an accomplished wood craftsman, photographer, dog
trainer, gourmet cook, teacher, raconteur, skier, naturalist, and all-
around legend in his home town of Aspen, Colorado. High above

Aspen, Stuart and his wife, Isabel, operated a shop called Toklat, which in Eskimo means "alpine headwaters," featuring an array of woodcrafts, Navajo rugs, jewelry, fish fossils, and photography. He would use his free time in summers to rebuild parts of a ghost town called Ashcroft for the U.S. Forest Service. He charged nothing for his time and labor. For groups venturing up the mountain from Aspen, he and Isabel would cook dinners featuring local foods cooked with style and simmered over great stories about the mountains, the town, and their lives. Stuart was seldom at a loss for words. His living, if that is an appropriate word for a how a Renaissance man earns his keep, was made as a woodworker. He and his sons crafted tables and cabinet-work with exquisite inlaid patterns using an assortment of woods from forests all over the world. A Mace table was like no other, and so was its price. Long before it was de rigueur to do so, Stuart bought his wood from forests managed for long-term ecological health. The calibration between ecological talk and do wasn't a thing for Stuart. He paid attention to details.

I first met Stuart in 1981. I was living in the Ozarks at the time and part of an educational organization that included, among other things, a farm and steam-powered sawmill. In the summer of 1981 one of our projects was to provide two tractor-trailer loads of oak beams for the Rocky Mountain Institute being built near Old Snowmass. Stuart advised us about cutting and handling large timber, about which we knew little. From that time forward Stuart and I would see each other several times a year either when he traveled through Arkansas or when I wandered into Aspen in search of relief from Arkansas summers. He taught me a great deal, not so much about wood per se as about the relation of ecology, economics, craftwork, generosity, and good-heartedness. I last saw Stuart in a hospital room shortly before he died of cancer in June 1993. In that final conversation, I recall Stuart being considerably less interested in the cancer that was consuming his body than in the behavior of the birds outside his window. He proceeded to deliver an impromptu lecture on the ecology of the Rocky Mountains. We cried a bit and hugged, and I went on my way. Shortly thereafter he went on his.

Every time I use his letter opener I think of Stuart. I believe that he intended it to be this way. For me the object itself is a lesson in giving and appropriate materialism. It is a useful thing. Hardly a day passes that I do not use it to open my mail, pry something open, or as a conversational aid to help emphasize a point. Second, it is beautiful. The coloring ranges from a deep brown to a tawny yellow. The wood is hard enough that it does not show much wear after a decade and a half of daily use. Third, it was made with great skill and design intelligence. The handle is carved to fit a right hand. Two fingers fit into a slight depression carved in the base. My thumb fits into another depression along the top of the shank. It is a pleasure to hold; its smoothness feels good to the touch. And it works as intended. The blade is curved slightly to the right, which serves to pull the envelope open as the blade slices through the paper.

Had Stuart been a typical consumer he could have saved himself some time and effort. He could have hurried to a discount office supply store to buy a cheap and durable chrome-plated metal letter opener stamped out by the tens of thousands in some third world country by underpaid and overworked laborers employed by a multinational corporation using materials carelessly ripped from the earth by another footloose conglomerate and shipped across the ocean in a freighter spewing Saudi crude every which way and sold by nameless employees to anonymous consumers in a shopping mall built on what was once prime farmland and is now uglier than sin itself making a few shekels for some organization that buys influence in Washington and seduces the public on TV. But you get the point.

In other words, had Stuart been a rational economic actor, he would have saved himself a lot of time that he could have used for watching the Home Shopping Channel. He could have maximized his gains and minimized his losses as the textbooks say he should do. Had he done so, he would have been participating in the great scam called the global economy, which means helping some third world country "develop" by selling the dignity of its people and their natural heritage for the benefit of others who lack for nothing. And he would have helped our own gross national product become all that much grosser.

A great global debate is under way about the sustainability and fairness of present patterns of consumption (Myers 1997, Sagoff 1997, Vincent and Panayotou 1997). On one side are those speaking for the poor of the world, various religious organizations, and the environment, who argue adamantly that wealthy Americans, Japanese, and Europeans consume far too much. Doing so, they believe, is unfair to the poor, future generations, and other species of life. This consumption is stressing the earth to the breaking point. Others, who believe themselves to be in the middle, argue it is not that we consume too much, only that we consume with too little efficiency. Below the surface of such views there is, I suspect, the gloomy conviction that short of an Ayatollah it is too late to reign in the hedonism loosed on the world by the advertisers and the corporate purveyors of fun and convenience. Human nature, they think, is inherently porcine, and given a choice, people wish only to see the world as an object to consume and the highest purpose of life to maximize bodily and psychological pleasure. For the managers, a better sort, a dose of more advanced technology and better organization will keep the goods coming. No problem. This view of human nature I take to be a self-fulfilling prophecy of the kind Dostoevsky's Grand Inquisitor would have appreciated. At the other end of the debate are the economic buccaneers and their sidekicks who talk glibly about more economic growth and global markets. A quick review of the seven deadly sins reveals them to be full-fledged heathens who will burn for eternity in hellfire. I know such things because I am the son of a Presbyterian preacher.

Because I believe that it is right and because I know it needs help, the first position in this debate is the one for which I intend to speak. I must begin by noting that "consume" as defined by the *New Shorter Oxford English Dictionary* means "destroy by or like fire or (formerly) disease." A "consumer," then, is "a person who squanders, destroys, or uses up." In this older and clearer view, consumption implied disorder, disease, and death. In our time, however, we proudly define ourselves not so much as citizens, or producers, or even as persons, but as consumers. We militantly defend our rights as consumers while letting our rights as citizens wither. Consumption is built into virtually everything we do. We have erected an economy, a society, and soon an entire planet around what was once recognized as a form of mental derangement. How could this have happened?

The emergence of the consumer society was neither inevitable nor accidental. Rather, it resulted from the convergence of a body of ideas that the earth is ours for the taking, the rise of modern capitalism, technological cleverness, and the extraordinary bounty of North America where the model of mass consumption first took root. More directly, our consumptive behavior is the result of seductive advertising, entrapment by easy credit, prices that do not tell the truth about the full costs of what we consume, ignorance about the hazardous content of much of what we consume, the breakdown of community, a disregard for the future, political corruption, and the atrophy of alternative means by which we might provision ourselves. The consumer society, furthermore, requires that human contact with nature, once direct, frequent, and intense, be mediated by technology and organization. In large numbers we moved indoors. A more contrived and controlled landscape replaced one that had been far less contrived and controllable. Wild animals, once regarded as teachers and companions, were increasingly replaced with animals bred for docility and dependence. Our sense of reality once shaped by our complex sensory interplay with the seasons, sky, forest, wildlife, savanna, desert, rivers, seas, and the night sky increasingly came to be shaped by technology and artificial realities. Urban blight, sprawl, disorder, and ugliness have become, all too often, the norm. Compulsive consumption, perhaps a form of grieving or perhaps evidence of mere boredom, is a response to the fact that we find ourselves exiles and strangers in a diminished world that we once called home.

Since stupidity is usually sufficient to explain what goes wrong in human affairs, a belief in conspiracies that require great cleverness is both superfluous and improbable. In this case, however, there is good reason to think that both were operative. Clearly we were naive enough to be suckered by folks like Lincoln Filene and Alfred Sloan who conspired to create a kind of human being that could be dependably exploited and even come to take a perverse pride in their servitude. The story has been told well by Thorstein Veblen (1973), Stuart Ewen (1976), William Leach (1993), and others and does not need to be repeated in detail here. In essence, it is a simple story. The first step involved bamboozling people into believing that who they are and what they owned were one and the same. The second step was to deprive people of alternative and often cooperative means by which they might provide basic needs and services. The destruction of light rail systems throughout the United States by General Motors and its co-conspirators, for example, had nothing to do with markets or public choices and everything to do with back-room deals designed to destroy competition with the automobile. The third step was to make as many people as possible compulsive and impulsive consumers, which is to say addicts, by the advertising equivalent of daily saturation bombing. The fourth step required giving the whole system legal standing through the purchase of several generations of politicians and lawyers. The final step was to get economists to give the benediction by announcing that greed and the pursuit of self-interest were, in fact, rational. By implication, thrift, a concern for others, public mindedness, farsightedness, or self-denial were old-fashioned and irrational. Add it all up and *Voila!* the consumer: an indoor, pleasure-seeking species adapted to artificial light, living on plastic money, and unable to distinguish the "real thing" (as in "Coca-Cola is . . .") from the real thing.

Do we consume too much? Certainly we do!

> Americans, who have the largest material requirements in the world, each directly or indirectly use an average of 125 pounds of material every day, or about 23 tons per year. . . . Americans waste more than 1 million pounds per person per year. This includes: 3.5 billion pounds of carpet sent to landfills, 25 billion pounds of carbon dioxide, and six billion pounds of polystyrene. Domestically, we waste 28 billion pounds of food, 300 billion pounds of organic and inorganic chemicals used for manufacturing and processing, and 700 billion pounds of hazardous waste generated by chemical production. . . . Total wastes, excluding wastewater, exceed 50 trillion pounds a year in the United States. . . . For every 100 pounds of product we manufacture in the United States, we create at least 3,200 pounds of waste. In a decade, we transform 500 trillion pounds of molecules into nonproductive solids, liquids, and gases. (Hawken 1997, 44)

Does compulsive consumption add to the quality of our lives? Beyond some modest level, the answer is no (Cobb et al. 1995). Does it satisfy our deepest longings? No, and neither is it intended to do so. To the contrary, the consumer economy is designed to multiply our dissatisfactions and dependencies. In psychologist Paul Wachtel's words: "Our present stress on growth and productivity is intimately related to the decline in rootedness. Faced with the loneliness and vulnerability that come with deprivation of a securely encompassing community, we have sought to quell the vulnerability through our possessions" (1983, 65). Do we feel guilty about the gluttony, avarice, greed, lust, pride, envy, and sloth that drive our addiction? A few may.

But most of us, I suspect, consume mindlessly and then feel burdened by having too much stuff. Our typical response is to hold a garage sale and take the proceeds to the mall and start all over again. Can the U.S. level of consumption be made sustainable for all 6.2 billion humans now on the earth? Not likely. By one estimate, to do so for just the present world population would require the resources of two additional planets the size of Earth (Wackernagel and Rees 1996).

If there ever was a bad deal, this is it. For a mess of pottage we surrendered a large part of our birthright of connectedness to each other and to the places in which we live, along with a sizable part of our practical competence, intelligence, health, community cohesion, peace of mind, and capacity for citizenship and neighborliness. Our children, consumers in training, can identify over a thousand corporate logos but only a dozen or so plants and animals native to their region. As a result they are at risk of living diminished, atomized lives. We consume, mostly in ignorance, chemicals like atrazine and alachlor in our cornflakes, formaldehyde in our plywood and particle board, and perchloroethylene in our dry-cleaned clothing (Fagin and Lavelle 1996). Several hundred other synthetic chemicals are embedded in our fatty tissues and circulate in our blood, with effects on our health and behavior that we will never fully understand. Our rural landscapes, once full of charm and health, are dying from overdevelopment, landfills, discarded junk, too many highways, too many mines and clear-cuts, and a lack of competent affection. Cities, where the civic arts, citizenship, and civility were born, have been ruined by the automobile. Death by overconsumption has become the demise of choice in the American way of life. The death certificates read "cancer," "obesity," and "heart disease." Some of our kids now kill each other over Nike shoes and jackets with NFL logos. Tens of thousands of us die on the highways each year trying to save time by consuming space. To protect our "right" to consume another country's oil, we have declared our willingness to incinerate the entire planet. We have, in short, created a culture that consumes everything in its path including its children's future. The consumer economy is a cheat and a fraud. It does not, indeed cannot, meet our most fundamental needs for belonging, solace, and authentic meaning.

"We must," in Wendell Berry's words, "daily break the body and shed the blood of creation. When we do this knowingly, lovingly, skillfully, reverently, it is a sacrament. When we do it ignorantly, greedily, clumsily, destructively, it is a desecration" (1981, 281). Can our use of the world be transformed from desecration to sacrament? Is it possible to create a society that lives within its ecological means, taking no more than it needs, replacing what it takes, depleting neither its natural capital nor its people, one that is ecologically sustainable and also humanly sustaining?

The general characteristics of that society are, by now, well known. First, a sustainable society would be powered by current sunlight, not ancient sunshine stored as fossil fuels. The price of an item in such a society would reflect, in Thoreau's words, "the amount of life which is required to be exchanged for it" (Thoreau 1971, 286), which is to say its full cost. This society would not merely recycle its waste but would eliminate the very concept of waste. Since "the first precaution of intelligent tinkering," as Aldo Leopold (1966, 190) once put it, "is to keep every cog and wheel," a sustainable society would hedge its bets by protecting both biological and cultural diversity. Such a society would exhibit the logic inherent in what is called "system dynamics" having to do with the way things fit together in harmonious patterns over long periods of time. Its laws, institutions, and customs would reflect an awareness of interrelatedness, exponential growth, feedback, time delays, surprise, and counterintuitive outcomes. It would be a smarter, more resilient, and ecologically more adept society than the one in which we now live. It would also be a more materialistic society in the sense that its citizens would value all materials too highly to treat them casually and carelessly. People in such a society would be educated to be more competent in making and repairing things and in growing their food. They would thereby understand the terms by which they are provisioned more fully than most of us do.

There is no good argument to be made against such a society. All the more reason to wonder why we have been so unimaginative and so begrudgingly slow to act on what later generations will see as merely an obvious convergence of prudent self-interest and ethics. It is certainly not for the lack of spilled ink, conferences in exotic places, and high-powered rhetoric. But sermons aiming to make us feel guilty about our consumption seldom strike a deep enough chord in most of us most of the time. The reason, I think, has to do with the fact that we are moved to act more often, more consistently, and more pro-

foundly by the experience of beauty in all of its forms than by intellectual arguments, abstract appeals to duty, or even by fear.

The problem is that we do not often see the true ugliness of the
consumer economy and so are not compelled to do much about it.
The distance between shopping malls and the mines, wells, corporate
farms, factories, toxic dumps, and landfills, sometimes half a world
away, dampens our perceptions that something is fundamentally
wrong. Even when visible to the eye, ugliness is concealed from our
minds by the very complicatedness of such systems which make it
difficult to discern cause and effect. It is veiled by a fog of abstract
numbers that measure our sins in parts per billion and as injustices
discounted over decades and centuries. It is cloaked by the ideology of
progress that transmutes our most egregious failures into chrome-
plated triumphs.

We have models, however, of a more transparent and comely
world beginning with better ways to provide our food, fiber, materials,
shelter, energy, and livelihood and to live in our landscapes. Over the
past 3.8 billion years, life has been designing strategies, materials, and
devices for living on earth. The result is a catalog of design wisdom
vastly superior to the best of the industrial age that might instruct us
in the creation of farms that function like prairies and forests, wastewater systems modeled after natural wetlands, buildings that accrue
natural capital like trees, manufacturing systems that mimic ecological processes, technologies with efficiencies that exceed those of our
best technologies by orders of magnitude, chemistry done safely with
great artistry, and economies that fit within their ecological limits
(Lyle 1994; Van der Ryn and Cowan 1996; Wann 1990). For discerning students, nature instructs about the boundaries and horizons of
our possibilities. It is the ultimate standard against which to measure
our use of the world.

The consumer economy was intended to liberate the individual
from community and material constraints and to thoroughly dominate nature and thereby to expand the human realm to its fullest.
Descartes, Galileo, Newton, Adam Smith, and their heirs, the architects of the modern world, assumed nature to be machinelike, with
no limits, and humans to be similarly machinelike, with no limits to
their wants. Consistent with those assumptions, excess has become
the defining characteristic of the modern economy, evidence of design failures that cause us to use too much fossil energy, too many

materials, and make more stuff than we could use well in a hundred lifetimes.

If, however, we intend to build durable and sustainable communities, and if we begin with the knowledge that the world is ecologically complex, that nature does in fact have limits, that our health and that of the natural world are indissolubly linked, that we need coherent communities, and that humans are capable of transcending their self-centeredness, a different design strategy emerges. For the design of a better society and healthier communities, in Vaclav Havel's words, "we must draw our standards from the natural world, heedless of ridicule, and reaffirm its denied validity. We must honour with the humility of the wise the bounds of that natural world and the mystery which lies beyond them, admitting that there is something in the order of being which evidently exceeds all our competence" (1987, 153).

Drawing our standards from the natural world requires that we first intend to act in ways that fit within larger patterns of harmony and health and create communities that fit within the natural limits of their regions. At a larger scale we must summon the political will to intend the creation of a civilization that calibrates the sum total of our actions with the larger cycles of the earth. When we do so, design at all scales entails not just the making of things, but becomes, rather, the larger artistry of making things that fit within their ecological, social, and historical context. Design is focused on rationality in its largest sense, giving priority to the wisdom of our intentions, not the cleverness of our means. Like the admonition to physicians to do no harm, the standard for ecological designers is to cause no ugliness, human or ecological, somewhere else or at some later time. When we get the design right, there is a multiplier effect which enhances the good order and harmony of the larger pattern. When we get it wrong, cost, disease, and disharmony multiply.

Like any applied discipline, ecological design has rules and standards. First, *ecological design is a community process that aims to increase local resilience by building connections between people, between people and the ecology of their places, and between people and their history*. The principle is an analog of engineering design, which aims to create resilience through redundancy and multiple pathways. Ecological design, similarly, works to counter the individualization, atomization, and dumbing-down inherent in the consumer economy by restoring connections at the community level. The process of design begins with questions such as, How does the proposed action fit the ecology of a place over time? Does it keep wealth within the community? Does it help people to become better neighbors and more competent persons? What are the true costs and who pays? What does it do for or to the prospects of our children and theirs?

Well-designed neighborhoods and communities are places where people need each other and must therefore resolve their differences, tolerate each other's idiosyncrasies, and on occasion, forgive each other. There is an architecture of connectedness that includes front porches facing onto streets, neighborhood parks, civic spaces, pedestrian-friendly streets, sidewalk cafes, and human scaled buildings (Jacobs 1961). There is an economy of connectedness that includes locally owned businesses that make, repair, and reuse, buying cooperatives, owner-operated farms, public markets, and urban gardens—patterns of livelihood that require detailed knowledge of the ecology of specific places. There is an ecology of connectedness evident in well-used landscapes, cultural and political barriers to the loss of ecologically valuable wetlands, forests, riparian corridors, and species habitat. Competent ecological design produces results tailored to fit the ecology of particular localities. There is an historical connectedness embedded in the memories that tie us to particular places, people, and traditions—swimming holes, lovers' lanes, campgrounds, forests, farm fields, beaches, ball fields, schools, historic sites, and burial grounds.

The degree to which connectedness now sounds distant from our present reality is a measure of how much we've lost in order to make consumption quick, cheap, and easy and to hide its true costs. Compulsive consumption is, in fact, proportional to the atomization of people, to social fragmentation, and to the emotional distance between people and their places. It is a measure of human incompetence requiring no skill and no wherewithal beyond ownership of a credit card. Connectedness, on the other hand, requires the ability to converse, to empathize, to resolve conflicts, to tolerate differences, to perform the duties of a citizen, to remember, and to re-member. It requires a knowledge of the natural history of a place, practical handiness, and place-specific skills and crafts. It creates roots, traditions, and a settled identity in a place.

Second, as described in chapter 4, *ecological design takes time seriously by placing limits on the velocity of materials, transportation, money, and information.* The old truism "haste makes waste" makes intuitively good ecological design sense. Increasing velocity often increases consumption, thereby generating more waste, disorder, and ugliness. In contrast, good design aims to use materials carefully and slowly. To preserve communities and personal sanity, it would place limits on the speed of transportation (Illich 1974). In order to take advantage of what economists call the "multiplier effect," it would slow the rate at which money is exchanged for goods and services imported from outside and thereby exits the local economy (Rocky Mountain Institute 1997). Good design aims to match the material requirements of the community with the clockspeed of charity and neighborliness, which is usually slower than that which is technologically feasible.

Excess consumption, in contrast, is in large measure relative to velocity. A bicycle, for example, moving at 20 miles per hour, requires only the energy of the biker. An automobile moving at 55 miles per hour for one hour will burn 2 gallons of gasoline. On a cross-Atlantic flight, a 747 flying at 550 miles per hour will burn 100 gallons of jet fuel per passenger. The difference is not just in the fuel consumed but also includes the entire support apparatus required by the increased speed of travel. A bicycle requires a relatively simple support infrastructure. An airline system, in contrast, requires a huge infrastructure including airports, roads, construction, manufacturing, and repair facilities, air-traffic control systems, mines, wells, refineries, banks, and the consumer industries that sell all of the paraphernalia of travel.

By taking time seriously enough to use it well, ecological design may also reset peoples' sense of propriety to a different moral time zone. The consumer society works best when people are impulsive buyers, expecting their gratifications instantly. By moderating the velocity of material flows, money, transport, and information, ecological design may also teach larger lessons having to do with the discipline of living within one's means, delaying gratification, the importance of thrift, and the virtue of nonpossessiveness.

Third, *ecological design eliminates the concept of waste and transforms our relationship to the material world.* The consumer economy uses and discards huge amounts of materials in landfills, air, and water. As a result, environmental policy is mostly a shell game that moves waste from one medium to another. Furthermore, carelessness in the making and using of materials has resulted in the global dissemination of some 100,000 synthetic chemicals carried by wind and water to the four corners of the earth.

Ecological design requires a higher order of competence in the making, use, and eventual reuse of materials than that evident in industrial economies. Ecologically, there is no such thing as waste. All materials are "food" for other processes. Ecological design is the art of linking materials in cycles and thereby preventing problems of careless use and disposal. Nature, accordingly, is the model for the making of materials. If nature did not make it, there are good evolutionary reasons to think that we should not. If we must, we ought to do so in small amounts that are carefully contained and biodegradable, which is to say, the way nature does chemistry. Nature makes living materials mostly from sunlight and carbon, and so should we. It does not mix elements like chlorine with mammalian biology. Neither should we. It creates novelty slowly, at a manageable scale, and so should we.

An economy that took design seriously would manage the flow of materials to maximize reuse, recycling, repair, and restoration. It would close waste loops by requiring manufacturers to take products back for disassembly and remanufacture. It would make distinctions between "products of service" and "products of consumption." In Europe, the concept is being applied to solvents, automobiles, and other products. In the United States, through the efforts of people like Ray Anderson and Bill McDonough, it is very slowly gaining acceptance.

Fourth, *ecological design at all levels has to do with system structure, not the rates of change.* The focus of ecological design is on systems and "patterns that connect" (Bateson 1979, 3–4). When we get the structure right, "the desired result will occur more or less automatically without further human intervention" (Ophuls 1992, 288). Consider two different approaches to the need for mobility. The Amish communities described in chapter 4 are structured around the capacity of the horse, which serves to limit human mischief, economic costs, consumption, dependence on the outside, and ecological damage, while providing time for human sociability, sources of fertilizer, and the peace of mind that comes with unhurriedness. In the Amish culture, the horse is a solar-powered, self-replicating, multifunctional structural solution that eliminates the need for continual management and regulation of people. Most of us are not about to become Amish, but we need to discover our own equivalent of the horse.

In the larger culture we expect laws and regulations to perform the same function, but they seldom do. The reason has to do with the fact that we tend to fiddle with particular symptoms rather than addressing structural causes of our problems. The Clean Air Act of 1970, for example, aimed to reduce pollution from auto emissions by attaching catalytic converters to each automobile—a coefficient solution. More than three decades later with more cars and more miles driven per car, even with lower pollution per vehicle, air quality is little improved and traffic is worse than ever. The true costs of that system include the health and ecological effects of air pollution and oil spills, the lives lost in traffic accidents, the degradation of communities, an estimated $300 billion per year in subsidies for cars, parking, and fuels, including the military costs of protecting our sources of imported oil, and the future costs of climate change. The result is a system that can only work expensively and destructively. A design solution to transportation, in contrast, would aim to change the structure of the system by reducing our dependence on the automobile through a combination of high-speed rail service, light-rail urban trains, bike trails, and smarter urban design that reduced the need for transportation in the first place.

The same logic applies to the structures by which we provision ourselves with food, energy, water, and materials and dispose of our waste. Much of our consumption, such as excessive packaging and preservatives in food, has been engineered into the system because of the requirements of long-distance transport. Some of our consumption is due to built-in obsolescence designed to promote yet more consumption. Some of it, such as the purchase of deadbolt locks and handguns, is necessary to offset the loss of community cohesion and trust caused in no small part by the culture of consumption. Some of our consumption is dictated by urban sprawl that leads to overdependence on automobiles. We have, in short, created vastly expensive and destructive structures to do what could be done better locally with far less expense and consumption. Redesigning such structures means learning how politics, tax codes, regulations, building codes, zoning, and laws work and how they might be made to work to promote ecological resilience and human sanity.

Without intending to do so, we have created a global culture of consumption that will come undone, perhaps in a few decades; perhaps it will take a bit longer. We are at risk of being engulfed in a flood of barbarism magnified by the ecologists' nightmare of overpopulation, resource scarcities, biotic impoverishment, famine, rampant disease, pollution, and climatic change. The only response that does credit to our self-proclaimed status as *Homo sapiens* is to rechart our course. That process, I believe, has already begun. But it will require far greater leadership, imagination, and wisdom to learn, and in some respects relearn, how to live in the world with ecological competence, technological elegance, and spiritual depth. We have models of communities, cultures, and civilizations that have in some measure done so and a few that continue to do so against long odds. There are still tribal people who know more than we will ever know about the flora and fauna of their places and who have over time created resource management systems that effectively limit consumption (Gadgil et al. 1993). There are sects, like the Amish, that continue to resist the consumer economy but nevertheless manage to live prosperous and satisfying lives. There are ancient practices, like Feng Shui, which has informed some of the best Chinese land use and architectural design for centuries, and new analytical skills such as least-cost, end-use analysis and geographic information systems that will help us see our way more clearly. There are also emerging interdisciplinary fields such as green architecture, restoration ecology, ecological engineering, solar design, sustainable agriculture, industrial ecology, and ecological economics that may in time come to constitute a full-fledged science of ecological design that may lay the foundations for a better world.

The problem is not one of potentials, but rather one of motivation. To live up to our potential we must first know that it is possible for us to live well without consuming the world's loveliness along with our children's legacy. But we must be inspired to act by examples that we can see, touch, and experience. Above all else, this is a challenge to educational institutions at all levels. We will need schools, colleges, and universities motivated by the vision of a higher order of beauty than that evident in the industrial world and that in prospect. They must help expand our ecological imagination and forge the practical and intellectual competence in the rising generation that turns merely wishful thinking into hopefulness.

Stuart's letter opener came to me as a gift, an embodiment of skill, design intelligence, kindness, and thrift. Stuart used no more than one-tenth of a board foot of wood to make it. He used no tools other than a wood rasp, some sandpaper, and linseed oil. The wood it-self was a product of sunlight and soil, symbolic of other and larger gifts. If I lose it, I will grieve, for it is full of memory and meaning. Each day I am reminded of Stuart and have a refresher course in the im-portance of craftsmanship, charity, and true economy. I will use it for a time and someday pass it on to another.

We gratefully acknowledge permission to reprint "The Ecology of Giving and Consuming," excerpted in somewhat altered form from *Consuming Desires: Consumption, Culture, and the Pursuit of Happiness*, ed. Roger Rosenblatt. Copyright © 1999 by Island Press. Reprinted by permission of Island Press/Shearwater Books, Washington, D.C., and Covelo, California. All rights reserved.

20

The Great Wilderness Debate, Again

Something will have gone out of us as a people if we ever let the remaining wilderness be destroyed; if we permit the last virgin forests to be turned into comic books and plastic cigarette cases; if we drive the few remaining members of the wild species into zoos or to extinction; if we pollute the last clear air and dirty the last clean streams and push our paved roads through the last of the silence, so that never again . . . can we have the chance to see ourselves single, separate, vertical, and individual in the world part of the environment of trees and rocks and soil, brother to the other animals, part of the natural world and competent to belong in it.

—Wallace Stegner

It is odd that attacks on the idea of wilderness have multiplied as the thing itself has all but vanished. Even alert sadists will at some point stop beating a dead horse. In the lower 48 states, federally designated wilderness accounts for only 1.8 percent of the total land area.

Including Alaskan wilderness, the total is only 4.6 percent. This is less than the land we've paved over for highways and parking lots. For perspective, Disney World is larger than one-third of our wilderness areas (Turner 1998, 619). Outside the United States there is little or no protection for the 11 percent of the earth that remains wild. It is to be expected that attacks on the last remaining wild areas would come from those with one predatory interest or another, but it is disconcerting that in the final minutes of the 11th hour they come from those who count themselves as environmentalists. Each of these critics claims to be for wilderness, but against the *idea* of wilderness. This fault line deserves careful scrutiny.[1]

In a recent article, for example, novelist Marilynne Robinson concludes that "we must surrender the idea of wilderness, accept the fact that the consequences of human presence in the world are universal and ineluctable, and invest our care and hope in civilization" (1998, 64). She arrives at this position not with joy, but with resignation. She describes her love of her native state of Idaho as an "unnameable yearning." But wilderness, however loved, "is where things can be hidden . . . things can be done that would be intolerable in a populous landscape." Has Robinson not been to New York, Los Angeles, Mexico City, or Calcutta, where intolerable things are the norm? But she continues: "The very idea of wilderness permits . . . those who have isolation at their disposal [to do] as they will" (ibid.). Presumably there would be no nuclear waste sites and no weapons laboratories without wilderness in which to hide them. She ignores the fact that the decisions to desecrate rural areas are mostly made by urban people or support one urban interest or another.

Robinson then comes to the recognition that history is not an uninterrupted triumphal march. There have been, she notes, a few dips along the way. The end of slavery in the United States produced a subsequent condition "very much resembling bondage" (Robinson 1998, 63). Now "those who are concerned about the world environment are the abolitionists of this era" whose "successes quite exactly resemble failure." So with a few successes under their belt, unnamed conservationists propose to establish a global "environmental policing system" and serve in the role of "missionary and schoolmaster" to the

1. The title of this chapter was borrowed from a book edited by Baird Callicott and Michael Nelson (1998).

rest of the world. But we cannot legitimately serve in that role because we, in the developed world, "have ransacked the world for these ornaments and privileges and we all know it" (ibid.). Accordingly, Robinson concludes that wilderness has "for a long time figured as an escape from civilization," so "we must surrender the idea of wilderness" (ibid., 64).

I have omitted some details, but her argument is clear enough. Robinson is against the idea of wilderness, but she does not tell us whether she is for or against preserving, say, the Bob Marshall or Gates of the Arctic, or whether she would give them away to AMAX or Mitsubishi. She is against the idea of wilderness because it seems to her that it has diverted our attention from the fact that "every environmental problem is a human problem" and we ought to solve human problems first. Whether environmental problems and human problems might be related, Robinson does not say.

The environmental movement certainly has its shortcomings. There are, in fact, good reasons to be suspicious of movements of any kind. But there is more at issue in Robinson's argument. The recognition that governments sometimes use less-populated areas for military purposes hardly constitutes a reason to fill up what's left of Idaho with shopping malls and freeways. Her assertion that abolition and environmentalism have produced ironic results is worth noting. But does she mean to say that we ought to ignore slavery, human rights abuses, toxic waste dumps, biotic impoverishment, or human actions that are changing the climate because we might otherwise incur unexpected and ironic consequences? Yes, rich countries have "ransacked the world," but virtually the only voices of protest have been those of conservationists aware of the limits of the earth. And what could she possibly mean by saying that "we are desperately in need of a new, chastened, self-distrusting vision of the world, an austere vision that can postpone the outdoor pleasures of cherishing exotica . . . and the debilitating pleasures of imagining that our own impulses are reliably good" (Robinson 1993, 64)? Are we to take no joy in the creation or find no solace or refuge in a few wild places? Who among us imagines their impulses to be reliably good? Would she confine us to shopping malls and a kind of indoor, air-conditioned introspection? Finally, Robinson seems not to have noticed that the same civilization in need of rehabilitation has done a poor job of protecting its land and natural endowment. Is it possible that human problems and environmental problems are reverse sides of the same coin of indifference and that we do not have the option of presuming to solve one without dealing with the other?

Robinson's broadside is only the latest salvo in a battle that began years earlier with articles by Ramachandra Guha (1998 a, 1998b), Baird Callicott (1991), and William Cronon (1995). The issues they raised were, to some extent, predictable. Guha, for example, believes that the designation of wilderness in many parts of the world has led to "the displacement and harsh treatment of the human communities who dwelt in these forests" (1998a, 273). His sensible conclusion is simply that "the export and expansion [of wilderness] must be done with caution, care, and above all, with humility" (ibid., 277).

Callicott's views and their subsequent restatement raise more complex and arcane issues. Callicott begins, as do most wilderness critics, by asserting that he is "as ardent an advocate" of wilderness as anyone and believes bird-watching to be "morally superior to dirtbiking" (1991, 339). The idea of wilderness may be wrong-headed, he thinks, "but there's nothing whatever wrong with the places that we call wilderness" (ibid., 587). He is discomforted by what he terms "the received concept of wilderness" inherited from our forebears who were all white males like Ralph Waldo Emerson, Henry David Thoreau, John Muir, Theodore Roosevelt, and Aldo Leopold. Callicott is unhappy with "what passes for civilization and its mechanical motif" that can conserve nature only by protecting a few fragments. He proposes, instead, to rescue civilization by "shift[ing] the burden of conservation from wilderness preservation to sustainable development" (ibid., 340). He proposes to "integrate wildlife sanctuaries into a broader philosophy of conservation that generalizes Leopold's vision of a mutually beneficial and mutually enhancing integration of the human economy with the economy of nature" (ibid., 346). This does not mean, however, "that we open the remaining wild remnants to development" (ibid.).

The heart of Callicott's argument, however, has to do with three deeper problems he finds in the idea of wilderness. Wilderness continues, he thinks, the division between humankind and nature. It is ethnocentric and causes us to overlook the effects tribal peoples had on the land. And, third, the very attempt to preserve wilderness is misplaced given the change characteristic of dynamic ecosystems. Callicott's critics, including philosopher Holmes Rolston, have responded by refuting these premises. Humans are not natural in the way Callicott supposes. There are "radical discontinuities between culture and nature" (Rolston 1991, 370). The 8 million or so tribal people living without horses, wheels, and metal axes had a relatively limited effect on the ecology of North America. After the initial colonization 10,000 or more years ago, the effects they did have, such as burning particular landscapes, did not differ much from natural disturbances such as fires ignited by lightning. As for the charge that conservationists are trying to preserve some idealized and unchanging landscape, Rolston asserts that "Callicott writes as if wilderness advocates had studied ecology and never heard of evolution. . . . Wilderness advocates do not seek to prevent natural change" (ibid., 375). To his critics, Callicott's dichotomy between wilderness preservation and sustainable development, as if these are mutually exclusive, makes little sense.

The dispute over wilderness went public in 1995 with the publication of William Cronon's essay "The Trouble with Wilderness, or Getting Back to the Wrong Nature" in the *New York Times Magazine*. Cronon did not add much that had not already been said, but he did give the debate a postmodern spin and the kind of visibility that lent considerable aid and comfort to the "wise use" movement and right-leaning opponents of wilderness. Remove the scholarly embellishments, and Cronon's piece is a long admonition to the effect that "we can(not) flee into a mythical wilderness to escape history and the obligation to take responsibility for our own actions that history inescapably entails. Most of all, it means practicing remembrance and gratitude for thanksgiving is the simplest and most basic of ways for us to recollect the nature, the culture, and the history that have come together to make the world as we know it" (1995a, 90).

Like Callicott, Cronon hopes that his readers understand that his criticism is "not directed at wild nature *per se* . . . but rather at the specific habits of thinking that flow from this complex cultural construction called wilderness" (1995a, 81). In other words, it is not "the things we label as wilderness that are the problem—for nonhuman nature and large tracts of the natural world *do* deserve protection—but rather what we ourselves mean when we use that label." That caveat notwithstanding, he proceeds to argue that "the trouble with wilderness is that it . . . reproduces the very values its devotees seek to reject." It represents a "flight from history" and "the false hope of an escape from responsibility." Wilderness is "very much the fantasy of people who have never themselves had to work the land to make a living" (ibid., 80). It "can offer no solution to the environmental and other problems that confront us." Instead, by "imagining that our true home is in the wilderness, we forgive ourselves the homes we actually inhabit" which poses a "serious threat to responsible environmentalism." The attention given to wilderness, according to Cronon, comes at the expense of environmental justice. Further, advocacy of wilderness "devalues productive labor and the very concrete knowledge that comes from working the land with one's own hands" (ibid., 85). But Cronon's principle objection is "that it may teach us to be dismissive or even contemptuous of . . . humble places and experiences," including our own homes.

Cronon concludes the essay by describing why the "cultural traditions of wilderness remain so important" (1995a, 88). He asserts that "wilderness gets us into trouble only if we imagine that this experience of wonder and otherness is limited to the remote corners of the planet, or that it somehow depends on pristine landscapes we ourselves do not inhabit" (ibid.). He admonishes us to pay attention to the wildness inherent in our own gardens, backyards, and landscapes.

"The Trouble with Wilderness" later appeared as the lead chapter in *Uncommon Ground: Toward Reinventing Nature* (Cronon 1995b). The authors' collective intention was to describe the many ways the concept of nature is socially constructed and to ask: "Can our concern for the environment survive our realization that its authority flows as much from human values as from anything in nature that might ground those values?" (ibid., 26). The book is a collage of the obvious, the fanciful, the "occulted,"[2] and disconnected postmodernism contrived as part of a University of California–Irvine conference titled "Reinventing Nature." The contributors were asked to summarize their thoughts in an addendum at the end of the volume titled "Toward a Conclusion," suggesting that they had not reached one. In an insightful retrospective, landscape architect Anne Whiston Spirn, author of the best chapter in the book, lamented the fact that the discussions were "so abstracted from the 'nature' in which we were liv-

---

2. The word is one used by Gary Snyder describing the same conference, "an odd exercise" he thought. See Gary Snyder, *A Place in Space*. (Washington: Counterpoint, 1995), p. 250.

ing . . . the talk seemed so disembodied" (ibid., 448). She wondered "how different our conversations might have been if they had not taken place under fluorescent lights, in a windowless room, against the whistling whoosh of the building's ventilation system" (ibid.). Indeed, the entire exercise of "reinventing nature" had the aroma of an indoor, academic, resume-building exercise. And the key assumption of the exercise—that nature can be reinvented—works only if one first conceives it as an ephemeral social construction. If nature is so unhitched from its moorings in hard physical realities, it can be recast as anything one fancies.

Not surprisingly, wilderness critics have received a great deal of criticism (Foreman 1994, 1996, 1998; Rolston 1991; Sessions 1995; Snyder 1995, 1996; Soule and Lease 1995; Willers 1996–1997). After the dust has settled a bit, what can be said of "the great new wilderness debate"? First, on the positive side, I think it can be said that, under provocation from Callicott, Cronon, and others, a stronger and more useful case for wilderness protection emerged (Foreman 1994, 1996, 1998; Grumbine 1996–1997; Noss 1998a, 1998b). The conjunction of older ideas about wilderness providing spiritual renewal and primitive recreation with newer ones concerning ecological restoration and the preservation of biodiversity offers a better and more scientifically grounded basis to protect and expand remaining wilderness areas in the twenty-first century. It is clear that we will need to fit the concept and the reality of wilderness into a larger concept of land use that includes wildlife corridors, sustainable development, and the mixed-use zones surrounding designated wilderness. But the origin of these ideas owes as much to Aldo Leopold as to any contemporary wilderness proponent. And, yes, environmentalists and academics alike need to make these ideas work for indigenous people, farmers, ranchers, and loggers. Development of conservation biology, low-impact forestry methods, and sustainable agriculture suggest that this is beginning to happen. For these advances, wilderness advocates can be grateful for their critics.

On a less positive note, the debate over wilderness resembles the internecine, hair-splitting squabbles of European socialists between 1850 and 1914. Often the differences between the various positions of that time were neither great nor consequential. Nonetheless, positions hardened, factions and parties formed around minutiae, and contentiousness and conspiracy became the norm on the political

Left. As a result, by 1914 the Left had coalesced into ideologically based factions, firmly and irrevocably committed to one impractical doctrine or another. It was a great tragedy that when the world needed far better ideas about the organization of property, government, and capital, in the early decades of the twentieth century, it had few from the Left. Instead, socialists of whatever stripe gave the strong impression to mainstream society that they had nothing coherent or reasonable to offer. Their language was obscure, their proposed solutions often entailed violence, their public manners were uncivil, and their tone was absolutist. It was in this environment that Lenin and his Bolsheviks concocted the odd brew of socialism, intolerance, brutality, messianic pretensions, and ancient czarist autocracy that became known as Marxism-Leninism. And the rest of the story, as they say, is history.

Like that of the early twentieth century, the world now more than ever needs better ideas about how to meld society, economy, and ecology into a coherent, fair, and sustainable whole. The question is whether environmentalists can offer practical, workable, and sensible ideas, not abstractions, arcane ideology, spurious dissent, and ideological hair-splitting reminiscent of nineteenth-century socialists. In this regard, the most striking aspect of the ongoing great wilderness debate is the similarity that exists between positions that were initially cast as mutually exclusive. There is no necessary divide, for example, between protecting wilderness and sustainable development. On the contrary, these are complementary ideas. And there are some issues, such as the old and unresolvable question about whether and to what degree humans are part of or separate from nature, that are hardly worth arguing about over and over again. Nor do we need to hear truisms that wilderness must be adapted to the circumstances, culture, and needs of particular places. These are obvious facts that deserve to be treated as such. Finally, since all participants profess support for the thing called wilderness, as distinct from the idea of it, we are entitled to ask, What is the point of the great wilderness debate? If we intend to influence our age in the little time we have, we must focus more clearly and effectively on the large battles that we dare not lose. The time and energy invested in our great debates should be judged against the sure knowledge that, while we argue among ourselves, others are busy bulldozing, clear-cutting, mining, building roads, and, above all, lobbying the powers that be.

Third, the effort to find common ground by "reinventing nature" along postmodernist lines seems to me to have the same foundational perspicacity as, say, the effort to extract sunbeams from cucumbers for subsequent use in inclement summers—a project of the great academy of Lagado, described by Jonathan Swift. Most surely we see nature through the lens of culture, class, and circumstance. Even so, it is remarkable how similarly nature is, in fact, "constructed" across different classes, cultures, times, and circumstances. This is so because gravity, sunlight, geology, soils, animals, and the biogeochemical cycles of the earth are the hard physical realities in which we live, move, and have our being. We are free to describe them in different symbols and wrap them in different cultural frameworks, but we do not thereby diminish their reality.

The idea that we are free to reinvent nature is, I think, an indulgence made possible because we have temporarily created an artificial world based on the extravagant use of fossil fuels. But that idea will not be particularly useful for helping us create a sustainable and sustaining civilization, however useful it may be as a reason to organize conferences in exotic places and for keeping postmodernists employed at high-paying, indoor jobs. "Reckless deconstructionism," in the words of Peter Coates, "cuts the ground from under the argument for the preservation of endangered species" (1998, 185). More broadly, it prevents us from taking any constructive action whatsoever. The postmodern contribution to environmentalism has privileged (in their word) an arcane, indoor, and ivory tower kind of environmentalism with more than a passing similarity to views otherwise found only on the extreme political right. Separated as it is from both physical and political realities, as well as the folks down at the truck stop, postmodernism provides no realistic foundation for a workable or intellectually robust environmentalism.

Looking ahead to the twenty-first century, the debate over wilderness has illuminated the fact that we will need larger, not smaller, ideas about land, nature, and ourselves. We will need more, not less, ecological imagination. We certainly need to be mindful of the "otherness" in our backyards, as Bill Cronon reminds us, but that reminder is a small idea that comes at a time when we must cope with global problems of species extinction, climatic change, emerging diseases, and the breakdown of entire ecosystems. We need a larger view of land and landscape than is possible where "It's mine and I'll do with it as I damn well please" is the prevailing philosophy. As Aldo Leopold pointed out decades ago, we need well-kept farms and homeplaces, well-managed forests, *and* large wilderness areas. None of these needs to compete with any other. Of the four, wilderness protection is by far the hardest to achieve. It is a societal choice that requires an ecologically literate public, political leadership, economic interests with a long-term view, and above all, the humility necessary to place limits on what we do. Until we have created a more far-sighted culture, the conjunction of these forces will always be rare, fragile, and temporary.

The battle over wilderness will grow in coming decades as the pressures of population growth and alleged economic necessity mount. There will be, someday soon, urgent calls to undo the Wilderness Act of 1964 and release much of the land it now protects to mining, economic expansion, and recreation facilities. At the same time it is entirely possible that much of our affection for wilderness, rural areas, and wildness will decline if we continue to become a tamer and more indoor people. In *Brave New World* (1932), Aldous Huxley described the effort to "condition the masses to hate the country" while conditioning them "to love all country sports." This process is already well under way, and we are the less for it. As D. H. Lawrence put it: "Oh, what a catastrophe for man when he cut himself off from the rhythm of the year, from his unison with the sun and the earth. Oh, what a catastrophe, what a maiming of love when it was made a personal, merely personal feeling, taken away from the rising and setting of the sun, and cut off from the magical connection of the solstice and equinox. This is what is wrong with us. We are bleeding at the roots" (Bass 1996, 21).

In the century ahead, the battle over wilderness will become a part of a much larger struggle. We have entered a new wilderness of sorts, one of our own making, consisting of technology that will offer us a virtual reality (an oxymoron if there ever was one), fun, excitement, and convenience. Caught between the ugliness that accompanies ecological decline and the siren call of a phony reality cut off from soils, forests, wildlife, and each other, we will be hard pressed to maintain our sanity and the best parts of our humanity. The struggle for wilderness and wildness in all of its forms is no less than a struggle over what we are to make of ourselves. I believe we need more wilderness and wildness, not less. We need more wildlands, wildlife, wildlife corridors, mixed-use zones, wild and scenic rivers, and, even urban wilderness. But above all, we need people who know in their bones that these things are important because they are the substrate of our humanity and an anchor for our sanity.

21

Loving Children: The Political Economy of Design

We are shocked when violence erupts in schoolyards or when a six-year-old child kills another in cold blood. But the headlines, which sensationalize such tragedies, reveal only the tip of what appears to be a larger problem that, given our present priorities, will only intensify. Youthful violence is symptomatic of something much bigger evident in diffuse anger, despair, apathy, the erosion of ideals, and rising level of teen suicide (up three-fold since 1960). Nationwide, 17 percent of children are on Ritalin, a central nervous system stimulant. Adults often respond with rejection and hostility, making a bad problem worse. We hire more psychologists and sociologists to study our children and more counselors to advise them about issues such as "anger management." As a result there are libraries of information about childhood, child psychology, child health, child nutrition, child behavior, and dysfunctional families, much of it quite beside the point. Then in desperation we hire more police to lock children up. We are crossing into a new pattern of relations between the generations, and much depends on how well we understand what is happening, why it is happening, and what is to be done about it.

The deeper causes of this situation are not apparent in the daily headlines and news reports. Dysfunctional families, depression, youthful violence, and the rising use of chemicals to sedate children are symptoms of something larger. Without anyone saying as much and without anyone intending to do so, we have unwittingly begun to undermine the prospects of our children and, at some level, I believe that they know it. This essay is a meditation on the larger patterns of our time and their effects on children. My argument is that the normal difficulties of growing up are compounded, directly and indirectly, by the reigning set of assumptions, philosophies, ideologies, and even mythologies by which we organize our affairs and conduct the business of society—what was once called "political economy." The study of political economy began with Adam Smith and continued on through Marx to the present in the work of scholars such as Yale University political scientist Charles Lindblom. Due to academic specialization and diminished public involvement in politics and community life, the field has declined. As a result, we have increasing difficulty in discerning larger social, economic, and political causes of our problems and doing something constructive about them. This essay is an attempt, in effect, to connect the dots describing those larger patterns. The first section below reviews evidence about the intersection of childhood and political economy from many different perspectives. The second section is a more explicit rendering of the political economy of contemporary global capitalism. The third and final section sketches some of the alternative political and economic arrangements necessary to honor our children and protect future generations.

The Evidence

Environmental Contaminants

By one estimate the average young American carries at least 190 chlorinated organic chemicals in his or her fatty tissues and bloodstream and another 700 additional contaminants as yet uncharacterized. Nursing infants in their first year of life have a higher body burden of dioxin than the average 70-year-old man (Thornton 2000). Children are threatened by the air they breathe, the food they eat, the water they drink, many of the materials common to everyday use, and fabrics in the designer clothes they wear. We have subjected our children to a vast experiment in which their body chemistry is subjected to hundreds of chemicals for which we have no evolutionary experience. We have good reason to suspect that their ability to procreate is being threatened by dozens of commonly used chemicals that disrupt the normal working of the endocrine system. As a result, sperm counts are falling and incidences of reproductive disorders of various kinds are rising (Colborn et al. 1996). We have reason to believe that exposure to some kinds of chemicals can cause varying levels of damage to the brain and nervous system. We have, in short, every reason to believe that a century of promiscuous industrial chemistry is seriously affecting our children. And we have reason to believe that current trends, unless altered, will grow worse. The scientific evidence is compelling but is widely dismissed because of a kind of deep-seated denial and a mind-set that demands absolute proof of harm before remedial action can be taken. So instead of eliminating the problem, we quibble about the rate at which we can legally poison each other.

Much of the same can be said about exposure to heavy metals. Nearly a million children under the age of five still suffer from low-level lead poisoning ("Dumbing Down the Children" 2000, part 1). Half of all children in the United States have lead levels that impair reading abilities (National Public Radio 2000). Even after leaded gasoline was phased out, Americans still have "average body burdens of lead approximately 300 to 500 times those found in our prehistoric ancestors" ("Dumbing Down the Children" 2000, part 3). The problem is not that we do not know the effects of lead and other substances on the human mind and body, but that corporations have the power to control public policy long after evidence of harm is established beyond reasonable doubt (Kitman 2000).

Nutrition and Exercise

More children exhibit the effects of bad diet and lack of exercise than ever before. The average diet of children has deteriorated in this age of affluence and fast food. Of those under the age of 19, one-quarter are overweight or obese. The U.S. Surgeon General believes that the problem is epidemic: "We see a nation of young people seriously at risk of starting out obese and dooming themselves to the difficult task of overcoming a tough illness" (Critser 2000, 150). Children are bombarded with 10,000 advertisements each year hawking fatty and sugar-laden food. The problem with a junk food diet is not just obesity, but the long-term damage it does to the pancreas, kidneys, eyes, nerves, and heart. There is a national eating disorder fostered by the corporations that feed us. But the disorder is not evenly visited on children. It is most apparent among children from lower-class homes. The junk diet of fat-laden fast foods represents a kind of class warfare in which corporations prey on the gullible, the poor, and the defenseless.

The problem of diet is compounded by a decline in physical exercise. One expert estimates that amount of physical activity of the typical child has declined 75 percent since 1900 (Healy 1990, 171). Another study shows a sharp decline in the average time children between the ages of 3 and 12 spend outdoors from an average of 1 hour and 26 minutes per day in 1981 to 42 minutes in 1997 (Fishman 1999). Indeed, capitalism works best when children stay indoors in malls and in front of televisions or computer screens. It loses its access to the minds of the young when they discover pleasures that cannot be bought.

Information

The average young person watches television a little over four hours per day. They are bombarded daily with the most tawdry kinds of "entertainment" and advertisements. Corporations spend $2 billion each year targeted specifically on the young, intending to lure them into a life of unthinking consumption. The American Academy of Pediatrics estimates that by age 18 they will have seen 360,000 television advertisements and 200,000 violent acts ("TV Viewed as a Public Health Threat" 2001). We have no good way to estimate the cumulative impact of all this on the growing human mind, but we may reasonably surmise that television strongly affects what they know and what they pay attention to and what they can know and pay attention to. We have, by one estimate, more than 1,000 studies showing that "significant exposure to media violence increases the risk of aggressive behavior in certain children and adolescents, desensitizes them to violence and makes them believe that the world is a 'meaner and scarier place' than it is" (ibid.). Young people are probably less adept with language than previous generations. They are increasingly hooked on the Internet, so that some colleges have had to hire counselors to deal with the problem as an addiction. And what has not happened in all the TV and Internet watching? The list is a long one: healthy contact with adults, making friends, outdoor exercise, reading, contemplation, and creative activity.

Education

With growing numbers of dysfunctional families, schools are now expected to make up for what parents ought to do. At the same time, schools and colleges are under increasing financial pressures and have increasingly become places of commerce. Many children are now exposed to the blatant commercialization of Channel One during school time. Many are required to read text materials developed by corporations that celebrate the virtues of capitalism without acknowledgment of its vices. More and more they are educated to take proficiency tests, not to learn creatively and critically. While we talk about the importance of learning, public spending tells a different story. A city like Cleveland, with one of the worst urban school systems in the nation, can find hundreds of millions of dollars for a new football stadium used eight times a year, but not the money or the foresight to repair the leaking roofs of its public schools. Nationally, some 60 percent of our schools need repair (Healy 1998, 92). Young people are quick to comprehend adult priorities. Financial priorities in higher education are also skewed. Commerce is making deep inroads into the academy, and colleges and universities have become heavily dependent on corporate support. As a result, corporations have acquired unprecedented influence over whole departments and the evolution of entire disciplines (Press and Washburn 2000).

Technology

A rising percentage of young people now spend many hours each day on the Internet or playing video games. Signs of trouble are already apparent. Internet addiction is a serious and growing problem. One study has shown that even a few hours a week on-line caused a "deterioration of social and psychological life" and higher levels of depression and loneliness among otherwise normal people (Harman 1998). The mental disorientation is caused by overexposure to a contrived electronic reality. As the technology for simulation advances, we may expect that the young so exposed will find increasing difficulty in distinguishing the contrived from the real and in establishing deep emotional ties to anyone or anything or simply taking responsibility for their own actions.

In the not-too-distant future, researchers in artificial intelligence and robotics are planning to create self-replicating machines that will be more intelligent than humans. Evolution, they say, works by replacement of the inferior by the superior, and these researchers unabashedly regard themselves as the agents of evolution with a mandate to create the next stage of intelligent life. It is not at all far-fetched to think that such alien intelligence could well find humans, meaning our children and grandchildren, inconvenient (Joy 2000). This is no longer some distant science fiction, but the reality coming inexorably into view. It is entirely possible that the present directions of technological development will create a world of simulated reality that will be more real to some in the next generation than the world as actually experienced. It is also increasingly possible that advances in fields such as artificial intelligence will diminish what it means to be human.

Ecology/Climate

The numbers are staggering. In the United States alone, we lose more than a million acres each year to urban sprawl, parking lots, and roads. We continue to destroy tropical forests worldwide at a rate of 80,000 square miles per year (Leakey and Lewin 237). The rate that we are driving species extinct rivals that of the last great extinction spasm 65 million years ago. Oceans and virtually every ecosystem on the planet are now deteriorating due to human activity. The scientific evidence indicates that climatic change is happening more rapidly than thought possible even a few years ago. Biotic impoverishment, climatic change, and pollution are beginning to undo millions of years of evolution and with it the rightful heritage of our children.

Were we to look at the plight of children worldwide, despite a burgeoning global economy, the story in many places is much worse. In some cities it is now common to see street children with no known parents and no home other than the street. They are sometimes killed or persecuted by police and preyed upon by those who exploit them shamelessly. It is common for children in third world countries to be used in the labor force under sweatshop conditions making products for global corporations. In Africa, the Balkans, the Middle East, and Ireland, children are caught in the middle of the worst kind of savagery. The facts differ from place to place but only as variations on a common theme of abuse, neglect, exploitation, and an astonishing level of intergenerational incompetence.

It is ironic that adults do not like the children they are raising. By one accounting, only 37 percent of adults believe that today's youth will "make this country a better place." Two-thirds of the adults surveyed find young people rude, spoiled, violent, and irresponsible (Applebome 1997). Ninety percent believe that values are not being transmitted to the young. And only one in five believe it common to find parents who are good role models for their children. No doubt previous generations often regarded the young with skepticism. What is different now, according to the authors of this study, is the intensity of antagonism between the generations and the empirical evidence supporting it. Daniel Goleman, author of *Emotional Intelligence* (1995), estimates that American children have declined on some 40 indicators of emotional and social well-being (cited in Healy 1998, 174).

Perhaps I have exaggerated the problems and the prospects for our children are quite different than I have described. Maybe these problems are mostly unrelated and arise from different causes. As any reader of Charles Dickens knows, children in earlier times were sometimes badly treated and lived in harsh conditions. And children from affluent homes are certainly not exposed to many hardships characteristic of some earlier times. But the evidence, in its entirety, is so well documented and so pervasive that we cannot mistake the larger pattern without thoroughgoing self-deception. We are unwittingly undermining our children's physical health, mental health, connection to adults, sense of continuity with the past, connections to nature, the health of ecosystems, a sense of commonwealth, and hope for a decent future. But we have difficulty in seeing whole systems in a culture shaped so thoroughly by finance capital and narrow specialization. However bad the situation of children in the past, no generation ever has done, or could have done, such systematic violence to its progeny and their long-term prospects. Most would adamantly protest that they love their children and are working as hard as possible to make a good life for them, and I believe that most parents and adults fervently believe that they are doing so. But we are caught in a pattern of deep denial that begins by confusing genuine progress, a difficult thing to appraise, with what is simply easy to measure—economic growth. We confuse convenience and comfort with well-being, longevity with health, SAT scores with real intelligence, and a rising GNP with real wealth. We express our affection incompetently. Without anyone intending to do so, we have launched a raid on their future, stealing things not rightfully ours, leaving behind a legacy destruction and degradation—a kind of intergenerational scorched earth policy. But why?

Political Economy

The conditions in which children experience nature is in large part an artifact of political economy, which Michael M'Gonigle defines as "the study of society's way of organizing both economic production and political processes that affect it and are affected by it" (1999a, 125). Beginning with Adam Smith and later Karl Marx, the study of political economy has aimed "to uncover and explain what might be called the 'system dynamics' of a society's processes of economic and political self-maintenance" (ibid., 126). The political economy of the modern world, in this view, is organized around the pursuit of economic growth, a science presumed to be value neutral, and the institutions of the state and corporation. Its ideology is "high modernist," which according to political scientist James C. Scott means "a muscle-bound version of the self-confidence about scientific and technical progress, the expansion of production, the growing satisfaction of human needs, the mastery of nature (including human nature), and, above all, the rational design of social order commensurate with the scientific understanding of natural laws" (1998, 4).

The main features of modern political economy are well known, even if their effects on childhood are not. The first and most obvious feature of contemporary political economy is the belief in the importance of economic growth and material accumulation. One day the major political fault line in the twentieth century about whether growth was to be organized by markets or governments will be seen as a minor doctrinal quibble. Regardless of specifics, economic growth has become the central goal for virtually every national government. Election outcomes are now more than ever an artifact of short-term economic performance. A second feature of modern political economy is the centrality of the global corporation. We are now provisioned with food, energy, materials, entertainment, health, livelihood, information, shelter, and transport by global corporations that operate with little oversight. The economic scale of the largest corporations dwarfs all but the largest national economies. As a result, corporations dominate national politics and policy and, through relentless advertising, the modern worldview as well. A third component of contemporary political economy is a particular kind of science rooted in the thinking of Descartes, Galileo, Bacon, and Newton. That science presumes a separation of subject from object, humankind from nature, and fact from value. Its power derives from its ability to reduce the objects of inquiry to their component parts. Its great weakness has been its inability to associate the knowledge so gained into its larger ecological, social, cultural, and normative context.

Political economy organized on these three pillars has many collateral effects on children. First, a society organized around economic growth is one that is in constant turmoil. Austrian economist Joseph Schumpeter (1978, 21–26) described the process by which physical capital is rendered obsolete as "creative destruction." Economic growth, then, means that the old and familiar is continually being replaced with something new and more profitable to the owners of capital. Similarly, the growth economy and the continual battle for market share among corporations is driven by and in turn drives a process of incessant technological change aiming for greater efficiency and speed. Creative destruction and technological dynamism, in turn, increase the velocity of lived experience. Not only is rapid change regarded as good, but rapid movement is as well. Corporations not only sell things, they sell sensation, movement, and speed, and these, too, are integral to the growth economy.

Little attention has been given to the effects of creative destruction, technological change, and increased velocity on the development of children, but they cannot be insignificant. For one thing, familiar surroundings and places where the child's psyche is formed are subject to continual modification, called "development," but to the child this is a kind of obliteration. But these places, regarded as real estate to the capitalist mind, are the places where children form their initial impressions of the world. Such places are, as Paul Shepard (1976) noted, the substrate for the adult mind. Some part of otherwise inexplicable teenage behavior in recent decades may be a kind of submerged grieving over the loss of familiar places rendered into housing tracts or shopping malls (Windle 1994). The effects of technological change and the consequent increase in the speed of lived experience on children is largely unknown, but it is reasonable to think that the healthy pace of human maturation is much slower than the frenetic speed of a technological society. The problem of speed is, I think, pervasive. At one level exposure to television (averaging more than four hours per person per day) with constantly changing images effects the neural organization of the mind in ways we do not understand. At another level, the decline in time spent with children means that parenting is compressed into smaller and smaller chunks of time. In either case, the child's sense of time is bent to fit technological and economic imperatives.

A second collateral effect arises from rampant materialism inherent in the growth economy. Childhood lived in more austere times was no doubt experienced differently from one lived in seemingly endless abundance. From birth on, children in an affluent culture marinate in a surfeit of things as well as the desire for things not yet possessed. Love in the growth economy is increasingly expressed by giving gifts, not by spending time with a child. Again, we have little idea of the long-term effects of excessive materialism on children, but it is reasonable to think that its hallmarks are satiation and shallowness and the loss of deeper feelings having to do with a secure and stable identity rooted in the self, relationships, and place. The important fact is not simply the effects of materialism but the more complex effects of the worldview conveyed in relentless advertising that hawks the message of instant gratification in a world of endless abundance. Whatever its other effects on the child, nature in a culture so lived can only recede in importance. Time once spent doing farm chores, exploring nearby places, fishing, or simply playing in a vacant lot has been replaced by the desire to possess or to experience some bought thing. It is, again, not far-fetched to think that one consequence is a loosening of ancient ties to place and an acquaintance with wildness. Nor is it unreasonable to suppose that the effect of several decades of glorifying money and things is now apparent in polls showing that the young increasingly want to get rich rather than live a life of deeper purpose.

A third collateral effect of contemporary political economy is that the world is increasingly rendered into commodities to be sold. Indeed, this is the purpose of the growth economy. Having saturated the market for automobiles and washing machines, it proceeded to sell us televisions and stereo equipment. Having saturated those markets, it moved on to sell us computers and cell phones. Eventually, it will sell us its version of reality that will be aimed to supplant more than most of us care to admit. Commodification, too, has its effects on the ecology of childhood. Those things that people once did for themselves as competent citizens or as self-reliant communities are now conveniently purchased. What's good for the gross national product, however, is often detrimental to communities. Real community can only be formed around mutual need, cooperation, sharing, and the daily exercise of practical competence. The effect of the growth economy and corporate dominance is to undermine the practical basis for community and with it the lineaments of trust. The absence of these qualities cannot be seen and so cannot be easily measured. Nonetheless, by many accounts there is a marked decline in community strength and social trust that cannot leave childhood unaffected (Putnam 2000). I suspect that these are mostly manifest in a decline in the imagination of a world of rich social possibilities that can only be lived out in real communities by people who have learned to live in interaction, not isolation. Instead, the young are socialized into an increasingly atomized world of extreme individualism governed by the assertion of freedoms without responsibilities. As such they are being trained to become reliable, even exuberant, consumers, but inept citizens and community members.

Much of the same can be said about the effects of economic growth on child care and the evolution of emotionally grounded intelligence in children. Economic necessity often forces both parents to work, leaving less time with their children. In psychiatrist Stanley Greenspan's words, one result of these social adaptations to economic forces is that "our nation has . . . launched a vast social experiment . . . and the early data are not encouraging" (1997, 179). What's at risk, he believes, are the "relationships on which developmental patterns rest" in a society in which "intimate personal interaction is declining and impersonality is increasing" (ibid., 169) These relationships, however, are crucial for the development of emotionally grounded intelligence.

Fourth, contemporary political economy is rooted in the tacit acceptance of high levels of risk that both jeopardizes the lives of children and colors their worldview. The growth economy creates mountains of waste, much of it toxic and some of it radioactive. This waste has been the driving force behind biotic impoverishment and the loss of biological diversity. Its further expansion now threatens climatic stability. Risks from technology and the scale of the economy are now pervasive, global, and permanent (Beck 1992). But the response of mainstream science, reflected in the practices of cost-benefit analysis or risk analysis, is rooted in the same kind of thinking that created the problems in the first place (O'Brien 2000). We have no way to know the full range of biophysical effects on children, nor can we say with certainty how they perceive the tapestry of risk that shrouds their future. But again, it is reasonable to think that these risks contribute to an undertone of despair and hopelessness.

Finally, the role of science in this larger political economy resembles more and more what Wendell Berry calls "modern superstition," in which "legitimate faith in scientific methodology seems to veer off into a kind of religious faith in the power of science to know all things and solve all problems" (2000, 18). Increasingly children grow up in a thoroughly secular culture, often without awareness that life is both gift and mystery. They are, in other words, spiritually impoverished. Because humans cannot live without meaning, the result is that their search for meaning, bereft of the possibility for authentic expression, can take ever more bizarre and futile forms.

It is certainly true that the situation of some children has improved vastly over what it was in the early years of capitalism when child labor was common. A full reading of the evidence, however, suggests caution in extrapolating too much. Improved living circumstances for some children fortunate enough to be raised in middle- or upper-class homes is a reality, with all of the caveats noted above. But little in contemporary political economy mandates that incomes will be fairly distributed or that children in other cultures will not be exploited to produce cheap sneakers and designer jeans for those living in affluence. Nor does this political economy afford adequate protection for any child living in the future from pollution, reproductive disorders, overexploitation of resources, climatic change, or loss of species.

Relative to their relation to nature, the reigning political economy has shifted the lives and prospects of children from:

- direct contact with nature to an increasingly abstract and symbolic nature
- routine and daily contact with animals to contact with man-made things
- immersion in community to isolated individualism
- less violence to more (much of it vicarious)
- direct exposure to reality to abstraction/virtual reality
- relatively slow to fast.

There are certainly exceptions. The Amish, for example, are notable because they are exceptions. On balance children in modern society are heavily shaped by a contemporary political economy that stresses materialism, economic growth, human domination of nature, and is tolerant of large-scale ecological risks with irreversible consequences. Their view of nature is increasingly distant, abstract, and utilitarian. However affluent, their lives are impoverished by diminishing contact with nature. Their imaginations, simulated by television and computers, are being impoverished ecologically, socially, and spiritually. The young, in Neil Postman's words, have been rendered into an "economic category . . . an economic creature, whose sense of worth is to be founded entirely on his or her capacity to secure material benefits, and whose purpose is to fuel a market economy" (Postman 2000, 125–126). This is not happening according to any plan; it is, rather, the logical outcome of the regnant system of political economy.

We have, in other words, created a global system of political economy in which it is not possible to be faithful or effective stewards of our children's future. It is a system that, by its nature, clogs many of its children's arteries with fast food. It is a system that, by its nature, poisons all of its children, albeit unevenly, with chemicals and heavy metals. It is a system that, by its nature, must saturate most of their minds with television advertisements and electronic trash. It is a system that, by its nature, must impoverish ecosystems and change climate. It is a system that, by its nature, undermines communities and family ties. It is a system, run by people who love their children, which will measure risks to them with great precision but is incapable, as it is, of implementing alternatives to those risks. It is a system that must remove most children from direct contact with unmanaged nature. And it is a system that encourages people to see the problems that arise from its very nature as anomalies, not as parts of a larger and deeply embedded pattern. We have unwittingly created a global political economy that prizes economic growth and accumulation of things above the well-being of children.

The important issues for our children are not narrowly scientific. They have little to do with symptoms and everything to do with systems. What kind of changes in the system of political economy would be necessary to protect the rights and dignity of children now and in the future?

A Child-Centered World

On July 30, 1998, the Supreme Court of the Philippines in *Minors Oposa* ruled that a group of 44 children had standing to sue on behalf of subsequent generations. In their suit, the children were trying to cancel agreements between timber companies and the Philippines government. The court found "no difficulty in ruling that they can, for themselves, for others of their generation and for the succeeding generations file a class suit . . . based on the concept of intergenerational responsibility insofar as the right to a balanced and healthful ecology is concerned" (quoted in Gates 2000, 289; see also Ledewitz 1998). The court considered the essence of that right to be the preservation of "the rhythm and harmony of nature" including "the judicious disposition, utilization, management, renewal and conservation of the country's forest, mineral, land, waters, fisheries, wildlife, off-shore areas and other natural resources" (Ledewitz 1998, 605). The court further stated that every generation has a responsibility to the next to preserve that rhythm and harmony for the full enjoyment of a balanced and healthful ecology." That right, the court argued, "belongs to a category . . . which may even predate all governments and constitutions . . . exist[ing] from the inception of humankind." Without the protection of such rights "those to come inherit nothing but parched earth incapable of sustaining life" (ibid.).

The court's decision recognizes what is, I think, simply obvious: that the right to a balanced and healthful ecology is the sine qua non for all other rights. The court acknowledged, in other words, that human health and well-being is inseparable from that of the larger systems on which we are utterly dependent. The court's decision implicitly acknowledges the inverse principle that no generation has a right to disrupt the biogeochemical conditions of the earth or to impair the stability, integrity, and beauty of biotic systems, the consequences of which would fall on subsequent generations as a form of irrevocable intergenerational remote tyranny.

No mention of ecological rights was made in our own Bill of Rights and subsequent constitutional development because, until recently, only the most prescient realized that we could damage the earth enough to threaten all life and all rights. But the idea that rights extend across generations was part of the revolutionary ethos of the late eighteenth century. The Virginia Bill of Rights (June 12, 1776), for example, held that "all men . . . have certain inherent rights, of which when they enter into a state of society, *they cannot by any compact deprive or divest their posterity*; namely, the enjoyment of life and liberty, with the means of acquiring and possessing property, and pursuing and obtaining happiness and safety" (emphasis added; quoted in Commager 1963, 103). That same idea was central to Thomas Jefferson's political philosophy. In the famous exchange of letters with James Madison in 1789, Jefferson argued that "the earth belongs in usufruct to the living . . . no man can, by natural right, oblige the lands he occupied, or the persons who succeed him in that occupation, to the paiment of debts contracted by him. For if he could, he might, during his own life, eat up the usufruct of the lands for several generations to come, and then the lands would belong to the dead, and not to the living" (Jefferson 1975, 445). Jefferson's use of the word "usufruct," the legal right of using and enjoying the fruits or profits of something belonging to another, is central to his point. For Jefferson, "the essence of the relationship between humans and the earth," in Richard Matthews's words, is "that of a trust, a guardianship, where the future takes priority over the present or past" (1995, 256). Initially skeptical, Madison, in time, came to hold a similar view (ibid., 260). On the other side of the political spectrum, Edmund Burke, the founder of modern conservatism, arrived at a similar position. In his *Reflections on the Revolution in France* ([1790] 1986), Burke described the intergenerational obligation to pass on liberties "as an entailed inheritance derived to us from our forefathers, and to be transmitted to our posterity" (119). For Burke, society is "a partnership not only between those who are living, but between those who are living, those who are dead, and those who are to be born" (ibid., 195).

It is reasonable, given what we now know, to enlarge the concept of intergenerational debt to include intergenerational ecological debts including biotic impoverishment, soil loss, ugly and toxic landscapes, and unstable climate. It is entirely logical to believe that the right to life and liberty presumes that the bearers of those rights also have prior rights to the biological and ecological conditions on which life and liberty depend. If Jefferson were alive now he would, I think, agree wholeheartedly with that amendment. Similarly, Burke would agree that the entailed inheritance of institutions, laws, and customs must also be expanded to include its ecological foundations without which there can be no useable inheritance at all. This suggests a convergence of Left and Right around the idea that the legitimate interests of our children and future generations sets boundaries to present behavior and changes the character of the present generation from property holders with absolute ecological rights to trustees for those yet to be born. The echo of this tradition is sounded in our time in documents such as the World Commission on Environment and Development report *Our Common Future*, which defines sustainable development as a way "to meet the needs and aspirations of the present without compromising the ability to meet those of the future" (1987, 40). Similarly, the "Earth Charter" aims, in part, to "transmit to future generations values, tradition, and institutions that support the long-term flourishing of Earth's human and ecological communities" (www.Earthcharter.org).

The extension of rights to some limits the freedom of others, thereby acknowledging that we live in a community and must be disciplined by the legitimate interests of every member of that community, now and in the future. Mesmerized by the industrial version of progress, we have been slow to recognize the revolutionary implications of that idea. But taken seriously, what do the ideas that children have standing to sue on behalf of the unborn or that certain ecological rights extend across time require of us? The answer is that we are required to follow the thread of obligations back to the economic and political conditions that affect children now and that will do so in the future. This requires, in short, that we rethink political economy from the perspective of those who cannot speak on their own behalf.

The most obvious of the present conditions affecting children has to do with the distribution of wealth. It is an article of faith in the contemporary political economy that everyone has the right to amass as much wealth as they possibly can and that any single generation has the same right vis-à-vis subsequent generations. As a result the top 1 percent in the United States have greater financial net worth than the remaining 95 percent (Gates 2000, 79). Working-class families have watched their real income decline by 7 percent between 1973 and 1998, putting more pressure on children who receive, as Jeff Gates puts it, "less parenting from substantially more stressed parents" (ibid., 47). Despite the huge increase in wealth in the past half-century, one-fifth of American children still live in poverty (ibid., 69). To guarantee that every child has the basics of food, shelter, medical care, decent parenting, and education means that we must address basic problems of economic security for families. Because poverty and its effects are often self-perpetuating across generations, inequity casts a long shadow over the future.

Similarly, implicit in the political economy of capitalism is the faith that the prosperity of the present generation will flow into the future as a positive stream of wealth. Losses in natural capital, it is assumed, will be offset by increased wealth. It is clear, however, that a stream of liabilities—toxic waste dumps, depleted landscapes, biotic impoverishment, climate change—cannot be nullified because natural and economic capital are not always interchangeable (Costanza and Daly 1992). The intergenerational balance of economic capital created minus the natural capital lost may not be positive because the costs of repairing, restoring, or simply adjusting to a world of depleted natural capital will exceed the benefits of advanced technology, sprawling cities, and larger stock portfolios.

Second, the recognition of children's rights would require us to rethink the taboo subject of property ownership. From that perspective we are obliged to protect not only the big components of the biosphere but also the small places in which children live. Children need access to safe places, parks, and wild areas. This recognition would cause us more often to rebuild decaying urban areas, restore degraded places, preserve more open spaces and river corridors, build more parks, set limits to urban sprawl, and repair ruined industrial landscapes. But doing so would require changing our belief in the nearly absolute rights of the landowner supposedly derived from English philosopher John Locke. We need to reread John Locke with the interests of children and future generations in mind. In fact, Locke's case for private ownership carried the caveat that land ownership should be limited so that "there is enough and as good left in common for others" (Locke [1688] 1965, 329; see also Schrader-Frechette 1993). The rights of children and future generations run counter to notions of property which give present owners the rights to do with land much as they please. At its most egregious, absentee corporations own land and subsurface mineral rights to large portions of Appalachia while paying minuscule taxes and practicing a kind of mining that decapitates entire mountains (Lockard 1998). Nothing in the law or current business ethics or mainstream economics would require them to give the slightest heed to the rights of the children living in those places or to those who will live there. Property rights, in a child-centered political economy, will require that owners must leave "enough and as good" or forfeit ownership.

Third, what do the rights of children mean for the interpretation of other rights such as the First Amendment guarantee of freedom speech and press? From a child's point of view that freedom has been corrupted to allow corporations to target children through advertising, movies, and television programming. More fundamentally, it has been corrupted to protect the rights of property, not the rights of people, by allowing corporations the same legal standing as persons. A child-centered political economy would, I think, permit no such reading of the Constitution or violations of common sense. Freedom of speech was intended by the founders, not as a license, but as a fundamental protection of religious and political freedoms and should not be interpreted as a right to prey on children for any purpose whatsoever.

Perhaps most difficult of all, what do the rights of children mean for the development of technology? Neil Postman once asked whether "a culture [could] preserve humane values and create new ones by allowing modern technology the fullest possible authority to control its destiny" (1982, 145). We have good reason to believe that the answer is no. But the subject is virtually taboo in the United States. Biologist Robert Sinsheimer (1978, 33) once proposed to limit the rights of scientists where their freedom to investigate was "incompatible with the maintenance of other freedoms." His argument was met with a thundering silence. In a society much enamored of invention, he inconveniently asked whether the rights of the inventor to create risky and dangerous technologies exceeded the rights of society to a safe and humane environment. Nearly a quarter of a century later, computer software engineer Bill Joy raised the same question regarding the rapid advance in technologies with self-replicating potential like genetic engineering, nanotechnologies, and robotics. In Joy's words, "we are being propelled into this new century with no plan, no control, no brakes" (2000, 256). Like Sinsheimer, Joy proposed placing limits on the freedom to innovate, assuming that the rights of some to pursue wealth, fame, or simply their curiosity should not trump the rights of future generations to a decent and humane world. A child-centered political economy would begin with the right of the child and future generations, not with those of the scientist and inventor. It would put brakes on the rights of technological change and scientific research where those might incur large and irreversible risks.

Fifth, a child-centered political economy would give priority to democratically controlled communities over the rights of finance capital and corporations—another taboo subject. In a series of decisions beginning with the *Dartmouth College* case and culminating in the 1886 *Santa Clara* case, the U.S. Supreme Court gave corporations the same protections given to individuals:

> We live in the shadow of a super-species, a quasi-legal organism that competes with humans and other life-forms in order to grow and thrive. . . . It can "live" in many places simultaneously. It can change its body at will—shed an arm or a leg or even a head without harm. It can morph into a variety of new forms absorb other members of its species, or be absorbed itself. Most astoundingly, it can live forever. To remain alive, it only needs to meet one condition: its income must exceed its expenditures over the long run. (Lasn and Liacas 2000, 41)

Corporations now rival or exceed the power and influence of nation-states. The largest 100 control 33 percent of the world's assets but employ only 1 percent of the world's labor (ibid.). They control trade, communications, agriculture, food processing, genetic materials, entertainment, housing, health care, transportation, and, not least, the political process. If there is anything left out of their control, it is because it is not profitable. Some routinely lie, steal, corrupt, and violate environmental laws with near impunity. As a consequence there is no safe future for children, nor are there safe communities in a world dominated by organizations that exist partly beyond the reach of law and owing no loyalty to anyone or to any place. The solutions are obvious. Corporations are chartered by the state and they can be dissolved by the state for just cause. We have implemented a "three strikes and you are out" standard for criminals; why not hold corporations and the people who serve them to the same standard? Wayne township in Pennsylvania, for example, bars any corporation with three or more regulatory violations within seven years. Many are asking for community control of investment capital and major assets. Nine midwestern states forbid corporate farm ownership. What attorney Michael Shuman (1998) calls "going local" requires a rejuvenation of democracy beginning by establishing local control over resources and investment decisions.

Finally, as farsighted and revolutionary as the decision of the Philippine court is, there is another and collateral right to be preserved, which is children's capacity to affiliate with nature and the places in which they live. Biologist Hugh Iltis describes that capacity thus: "Our eyes and ears, noses, brains, and bodies have all been shaped by nature. Would it not then be incredible indeed, if savannas and forest groves, flowers and animals, the multiplicity of environmental components to which our bodies were originally shaped, were not, at the very least, still important to us?" (quoted in Shepard 1998, 136). Harvard biologist E. O. Wilson calls this capacity "biophilia," which he defines as "the urge to affiliate with other forms of life" (1984, 85). "We are a biological species and will find little ultimate meaning apart from the remainder of life" (ibid., 81). Rachel Carson defined this capacity simply as "the sense of wonder" aided and abetted by "the companionship of at least one adult" (Carson [1956] 1984, 45).

Is the opportunity to develop biophilia and a sense of wonder important? Can it be considered a right? The answer to the first question is yes, because it is unlikely that we will want to preserve nature only for utilitarian reasons. We are likely to save only what we have first come to love. Without that affection, we are unlikely to care about the destruction of forests, the decline of biological diversity, or the destabilization of climate. To the second question the answer must again be affirmative because affiliation with nature, by whatever name, is an essential part of what makes us human. We have good reason to believe that human intelligence evolved in direct contact with animals, landscapes, wetlands, deserts, forests, night skies, seas, and rivers. We have reason to believe that "the potential for becoming as fully intelligent and mature as possible can be hindered and even mutilated by circumstances in which human congestion and ecological destitution limit the scope of experience" (Shepard 1998, 127). We can all agree that the act of deliberately crippling a child would violate basic rights. By the same token, mutilation of a child's capacity to form what theologian Thomas Berry (2000, 15) calls "an intimate presence within a meaningful universe," although harder to discern, is no less appalling because it would deprive the child of a vital dimension of experience. According to Berry:

> We initiate our children into an economic order based on exploitation of the natural life systems of the planet. To achieve this attitude we must first make our children unfeeling in their relation with the natural world. . . . For children to live only in contact with concrete and steel and wires and wheels and machines and computers and plastics, to seldom experience any primordial reality or even to see the stars at night, is a soul deprivation that diminishes the deepest of their human experiences. (2000, 15, 82)

The result of that deprivation is a kind of emotional and spiritual blindness to the larger context in which we live, abridging the sense of life.

Were we to take the right to a balanced and healthful ecology seriously, we would do all in our power to protect the right of children to develop a healthy kinship with the earth. We would honor the ancient tug of the Pleistocene in our genes by preserving opportunities for children to "soak in a place and [for] the adolescent and adult . . . to return to that place to ponder the visible substrate of his or her own personality" (Shepard 1996, 106). We would "find ways to let children roam beyond the pavement, to gain access to vegetation and earth that allows them to tunnel, climb, or even fall" (Nabhan and Trimble 1994, 9). We would preserve the right to "the playful exploration of habitat . . . as well as the gradual accumulation of an oral tradition about the land [that] have been essential to child development for over a million years" (ibid., 83). We would preserve wildness even in urban settings. This is not nature education as commonly understood. It is, rather, a larger subject of how and how carefully we manage the ecology of particular places to permit the full flowering of human potentials.

Conclusion

The invention of childhood in the late Middle Ages was a discovery, of sorts, that children were not simply miniature adults but were in a distinct stage of life with its own needs and developmental pattern (Aries 1962). This was more than a useful discovery; it was a fundamental acknowledgment that a decent culture needed to make a greater effort to shelter, nourish, and establish individual personhood than had previously been the case. We have good evidence from many sources that childhood as a distinct and protected phase of life is disappearing, and we have every reason to fear that loss. The primary cause is an errant system of political economy loosed on the world. It is failing children now and will in time fail catastrophically. Children will bear the brunt of that failure as well. Far from having settled all of the big political and economic issues, we have yet to create a political economy that protects the biosphere and the physical, mental, emotional, and spiritual well being of children and through them the future of our species. I hope we are at the beginning of what Thomas Berry calls the Ecozoic era, "when humans will be present to the Earth in a mutually enhancing manner" (2000, 55). For that hope to become manifest, we must first organize our political and economic affairs in a way that honors the rights of all children. The irony of our situation is that what appears from our present vantage point to be altruism will, in time, come to be seen as merely practical, farsighted self-interest.

Bibliography

Abram, D. 1996. *The Spell of the Sensuous*. New York: Pantheon.

Aeschliman, Michael. 1983. *The Restitution of Man: C. S. Lewis and the Case against Scientism*. Grand Rapids, Mich.: Eerdmans.

Alexander, C., et al. 1977. *A Pattern Language*. New York: Oxford University Press.

Anderson, E. N. 1996. *Ecologies of the Heart*. New York: Oxford University Press.

Anderson, R. 1998. *Mid-Course Correction*. Atlanta: Peregrinzilla Press.

Appalachian Land Ownership Task Force. 1983. *Who Owns Appalachia?* Lexington: University Press of Kentucky.

Applebome, Peter. 1997. "Children Score Low in Adults' Esteem." *New York Times*, June 26, A25.

Aries, Phillipe. 1962. *Centuries of Childhood*. New York: Vintage.

Ausubel, J. 1996. "Liberation of the Environment." *Daedalus* 125, no. 3: 1–18.

Bacon, F. [1627] 1965. *Francis Bacon: A Selection of His Works*. Ed. Sidney Warhaft. New York: Odyssey Press.

Bailey, R. 1995. *The True State of the Planet*. New York: Free Press.

Banuri, T., and A. Marglin. 1993. *Who Will Save the Forests?* London: ZED.

Barlett, Donald, and James Steele. 1998. "Corporate Welfare." *Time*, November 9, 37–54.

Bass, Rick. 1996. *The Book of the Yaak*. Boston: Houghton-Mifflin.

Bates, Sarah, et al. 1993. *Searching Out the Headwaters*. Washington: Island Press.

Bateson, Gregory. 1979. *Mind and Nature*. New York: E. P. Dutton.

Beck, Ulrich. 1992. *Risk Society*. Beverly Hills, Calif.: Sage.

Benyus, Janine 1997. *Biomimicry: Innovation Inspired by Nature*. New York: William Morrow.

Berman, Daniel, and John T. O'Connor. 1996. *Who Owns the Sun?* White River Junction, Vt.: Chelsea Green.

Berman, M. 1989. *Coming to Our Senses*. New York: Simon and Schuster.

Berry, Thomas. 1999. *The Great Work*. New York: Bell Tower.

Berry, Wendell. 1977. *The Unsettling of America*. San Francisco: Sierra Club Books.

———. 1981. *The Gift of Good Land*. San Francisco: North Point Books.

———. 1983. *Standing by Words*. San Francisco: North Point Books.

———. 1999. "Back to the Land: The Radical Case for Local Economy." *Amicus* 20 (Winter): 37–40.

———. 2000. *Life Is a Miracle*. Washington: Counterpoint Press.

Betts, Kellyn S. 1998. "A New 'Green' Building on Campus." *Environmental Science & Technology* 32, no. 17 (September 1): 412–414.

Boorstin, D. [1961] 1978. *The Image*. New York: Atheneum.

Bornstein, D. 1998. "Changing the World on a Shoestring." *Atlantic Monthly*, January, 34–38.

Bowers, C. A. 1993. *Critical Essays on Education, Modernity, and the Recovery of the Ecological Imperative*. New York: Teachers' College Press.

———. 2000. *Let Them Eat Data*. Athens: University of Georgia Press.

Brand, S. 1999. *The Clock of the Long Now*. New York: Basic Books.

Burke, E., [1790] 1986. *Reflections on the Revolution in France*. New York: Penguin.

Butler, G. 1996. Remarks to the National Press Club, Washington, D.C., December 4.

———. 1998. Remarks to the National Press Club, Washington, D.C., February 2.

Caldwell, L., and K. Shrader-Frechette. 1993. *Policy for Land*. Lanham, Md.: Rowman & Littlefield.

Callicott, J. Baird. 1991a. "The Wilderness Idea Revisited: The Sustainable Development Alternative." *Environmental Professional* 13: 235–247.

———. 1991b. "That Good Old-Time Wilderness Religion." *Environmental Professional* 13: 378–379.

Callicott, J. Baird, and Michael Nelson, eds. 1998. *The Great New Wilderness Debate*. Athens: University of Georgia Press.

Carson, Rachel. 1962. *Silent Spring*. Boston: Houghton Mifflin.

———. [1956] 1984. *The Sense of Wonder*. New York: Harper & Row.

Chargaff, Erwin. 1980. "Knowledge without Wisdom." *Harper's*, May, 41–48.

Coates, Peter. 1998. *Nature*. Berkeley: University of California Press.

Cobb, Clifford, et al. 1995. "If the GDP Is Up, Why Is America Down?" *Atlantic Monthly*, October, 59–78.

Colborn, Theo, Dianne Dumanoski, and J. P. Meyers. 1996. *Our Stolen Future*. New York: Dutton.

Collins, T. 2001. "Toward Sustainable Chemistry." *Science* 291 (January 5): 48–49.

Commager, H. S. 1963. *Documents of American History*. New York: Appleton, Century, Crofts.

Conquest, R. 1999. *Reflections on a Ravaged Century*. New York: Norton.

Costanza, R., and H. Daly. 1992. "Natural Capital and Sustainable Development." *Conservation Biology* 6, no. 1 (March): 37–46.

Crevier, D. 1994. *AI*. New York: Basic Books.

Critser, Greg. 2000. "Let Them Eat Fat." *Harper's*, March, 41–47.

Cronon, William. 1995. "The Trouble with Wilderness; or, Getting Back to the Wrong Nature." In *Uncommon Ground: Toward Reinventing Nature*, pp. 69–90. New York: W. W. Norton.

Daly, H. 1996. *Beyond Growth*. Boston: Beacon Press.

Damasio, A. 1994. "Descartes' Error and the Future of Human Life." *Scientific American*, October, 144–149.

Davis, D. B. 1984. *Slavery and Human Progress*. New York: Oxford University Press.

Deloria, V. 1999. *For This Land*. London: Routledge.

Dobb, E. 1996. "Pennies from Hell." *Harper's*, October, 39–54.

Donahue, B. 1999. *Reclaiming the Commons*. New Haven, Conn.: Yale University Press.

"Dumbing Down the Children" (Parts 1–3). 2000. *Environment and Health Weekly* nos. 687 (February 17), 688 (February 24), and 689 (March 2).

Easterbrook, G. 1995. *A Moment on the Earth*. New York: Viking.

Edelman, M. 1962. *The Symbolic Uses of Politics*. Urbana-Champaign: University of Illinois Press.

Ehrenfeld, David. 1978. *The Arrogance of Humanism*. New York: Oxford University Press.

Ehrlich, P., G. Daily, S. Daily, N. Myers, and J. Salzman. 1997. "No Middle Way on the Environment." *Atlantic Monthly*, December, 98–104.

Ellul, J. 1980. *The Technological System*. New York: Continuum.

———. 1990. *The Technological Bluff*. Grand Rapids, Mich.: Eerdmans.

E Source. 1992. *Energy-Efficient Buildings: Institutional Barriers and Opportunities*. Boulder, Colo.: E Source.

Ewen, S. 1973. *Captains of Consciousness*. New York: McGraw-Hill.

Fagin, D., and M. Lavelle. 1996. *Toxic Deception*. Secaucus, N.J.: Birch Lane Press.

Finley, M. I. 1980. *Ancient Slavery and Modern Ideology*. New York: Viking.

Fishman, C. 1999. "The Smorgasbord Generation." *American Demographics*, 54–60.

Foreman, Dave. 1994. "Wilderness Areas Are Vital." *Wild Earth* 4, no. 4, pp. 64–68.

———. 1996. "All Kinds of Wilderness Foes." *Wild Earth* 6, no. 4, pp. 1–4.

———. 1998. "Wilderness Areas for Real." In *The Great New Wilderness Debate*, ed. J. Baird Callicott and Michael Nelson, pp. 395–407. Athens: University of Georgia Press.

Forrester, J. 1971. "Counter-Intuitive Behavior of Social Systems." *Technology Review* 73, no. 3 (January): 52–67.

Franklin, C. 1997. "Fostering Living Landscapes." In *Ecological Design and Planning*, ed. G. Thompson and F. Steiner. New York: Wiley & Sons.

Gadgil, M., et al. 1993. "Indigenous Knowledge for Biodiversity Conservation." *Ambio* 22, nos. 2–3 (May): 151–156.

Gates, Jeff. 1998. *The Owership Solution: Toward a Shared Capitalism for the 21st Century*. Reading, Mass.: Addison-Wesley.

———. 2000. *Democracy at Risk*. Cambridge, Mass.: Perseus.

Gelbspan, R. 1998. *The Heat Is On*, 2d ed. Reading, Mass.: Perseus Books.

Gladwin, T., W. Newburry, and E. Reiskin. 1997. "Why Is the Northern Elite Mind Biased against Community, the Environment, and Sustainable Development?" In *Environment, Ethics, and Behavior*, ed. M. Bazerman et al., pp. 234–274. San Francisco: New Lexington Press.

Goleman, Daniel. 1995. *Emotional Intelligence*. New York: Bantam.

Gowdy, J., and C. McDaniel. 1995. "One World, One Experiment." *Ecological Economics* 15: 181–192.

Greenspan, Stanley. 1997. *The Growth of the Mind*. Reading, Mass.: Perseus.

Grossman, R., and F. Adams. 1993. *Taking Care of Business*. Cambridge: Charter Ink.

Grumbine, Edward. 1996–1997. "Using Biodiversity as a Justification for Nature Protection." *WildEarth* 6, no. 4 (Winter): 71–80.

Guha, Ramachandra. 1998a. "Deep Ecology Revisited." In *The Great New Wilderness Debate*, ed. J. Baird Callicott and Michael Nelson, pp. 271–279. Athens: University of Georgia Press.

———. 1998b. "Radical American Environmentalism and Wilderness Preservation: A Third World Critique." In *The Great New Wilderness Debate*, ed. J. Baird Callicott and Michael Nelson, pp. 231–245. Athens: University of Georgia Press.

Hardin, G. 1968. "The Tragedy of the Commons." *Science* 162 (December 13): 8–13.

Harman, Amy. 1998. "Sad Lonely World Discovered in Cyberspace." *New York Times*, September 30, pp. 1, 22.

"Harper's Index." 2000. *Harper's*, August, 11.

Hassanein, N. 1999. *Changing the Way America Farms*. Lincoln: University of Nebraska Press.

Havel, V. 1987. *Living in Truth*. London: Faber and Faber.

———. 1991. *Disturbing the Peace*. New York: Vintage.

———. 1992. *Summer Meditations*. New York: Knopf.

Hawken, P. 1993. *The Ecology of Commerce*. New York: HarperCollins.

———. 1997. "Natural Capitalism." *Mother Jones*, April, 40–53.

Hawken, P., A. Lovins, and H. Lovins. 1999. *Natural Capitalism*. Boston: Little, Brown.

Hawkes, J. 1951. *A Land*. New York: Random House.

Healy, Jane. 1990. *Endangered Minds*. New York: Simon & Schuster.

———. 1998. *Failure to Connect*. New York: Simon & Schuster.

Herbert, B. 1995. "A Nation of Nitwits." *New York Times*, March 1, A15.

Heschel, A. J. [1951] 1990. *Man Is Not Alone: A Philosophy of Religion*. New York: Farrar, Straus, and Giroux.

Horwitz, Morton. 1992. *The Transformation of American Law, 1870–1960*. New York: Oxford University Press.

Houghton, J. 1997. *Global Warming: The Complete Briefing*, 2d ed. New York: Cambridge University Press.

Hunter, J. Robert. 1997. *Simple Things Won't Save the Earth*. Austin: University of Texas Press.

Huxley, Aldous. [1932] 1998. *Brave New World*. New York: Harper Perennial.

Hyde, L. 1983. *The Gift*. New York: Vintage.

Ikle, F. 1994. "Growth without End, Amen?" *National Review*, March 7, 36–44.

Illich, I. 1974. *Energy and Equity*. New York: Harper Perennial.

Intergovernmental Panel on Climate Change. 2001. *Climate Change 2001*. 3 vols. New York: Cambridge University Press.

Jackson, W. 1985. *New Roots for Agriculture*. Lincoln: University of Nebraska Press.

Jacobs, J. 1961. *The Life and Death of Great American Cities*. New York: Vintage.

James, W. 1955. *The Will to Believe and Other Essays*. New York: Dover.

Jefferson, T. 1816. Letter to Charles Yancey, January 6. *Bartlett's Familiar Quotations*, 14th ed.

———. 1975. *The Portable Thomas Jefferson*. Ed. M. Peterson. New York: Viking.

Joy, Bill. 2000. "Why the Future Doesn't Need Us." *Wired*, April, 238–262.

Kahn, H., and W. Brown. 1976. *The Next Two Hundred Years*. New York: William Morrow.

Kellert, S., and E. O. Wilson. 1993. *The Biophilia Hypothesis*. Washington: Island Press.

Kirk, R. 1982. *The Portable Conservative Reader*. New York: Penguin.

Kitman, Jamie. 2000. "The Secret History of Lead." *The Nation*, March 20, 11–44.

Kohr, Leopold. 1980. "Slow Is Beautiful," *Co-evolution Quarterly*, Summer, 57–59.

Kozol, J. 1985. *Literate America*. New York: Anchor-Doubleday.

Kuhn, Thomas. 1962. *The Structure of Scientific Revolutions*. Chicago: University of Chicago Press.

Kurzweil, R. 1999. *The Age of Spiritual Machines*. New York: Penguin.

Langer, S. [1942] 1976. *Philosophy in a New Key*. Cambridge: Harvard University Press.

Lansing, J. S. 1991. *Priests and Programmers*. Princeton, N.J.: Princeton University Press.

Lasn, Kalle, and Tom Liacas. 2000. "Corporate Crackdown." *Adbusters Magazine*, August–September, 38–49.

Leach, W. 1993. *Land of Desire*. New York: Pantheon.

Leakey, Richard, and Roger Lewin. 1995. *The Sixth Extinction*. New York: Doubleday.

Ledewitz, Bruce. 1998. "Establishing a Federal Constitutional Right to a Healthy Environment in U.S. and Our Posterity." *Mississippi Law Journal* 68, no. 2: 565–673.

Leopold, A. 1953. *Round River*. New York: Oxford University Press.

———. 1966. *A Sand County Almanac*. New York: Ballantine.

———. 1999. *The Essential Aldo Leopold*. Ed. C. Meine and R. Knight. Madison: University of Wisconsin Press.

Levins, Richard. 1998. "Looking at the Whole: Toward a Social Ecology of Health." Robert H. Ebert Lecture, Kansas Health Foundation.

Lewis, C. S. [1947] 1970. *The Abolition of Man*. New York: Macmillan.

Lewis, M. 1992. *Green Delusions*. Durham, N.C.: Duke University Press.

Lockard, Duane. 1998. *Coal: A Memoir and Critique*. Charlottesville: University Press of Virginia.

Locke, J. [1689] 1965. *Two Treatises of Government*. New York: Mentor Books.

Lovins, A. 1976. "The Road Not Taken?" *Foreign Affairs* 55, no. 1 (October): 65–95.

Lovins, A., and H. Lovins. 1982. *Brittle Power*. Andover, Mass.: Brick House.

Lovins, Amory, and Andre Lehmann. 1997. *Small Is Profitable*. Snowmass, Colo.: Rocky Mountain Institute.

Ludwig, D., R. Hilborn, and C. Walters. 1993. "Uncertainty, Resource Exploitation, and Conservation." *Science* 260 (April 2): 17, 36.

Lyle, J. 1994. *Regenerative Design for Sustainable Development*. New York: John Wiley.

———. 1997. "Green Infrastructure." Unpublished ms.

Martin, C. 1999. *The Way of the Human Being*. New Haven, Conn.: Yale University Press.

Maslow, A. 1966. *The Psychology of Science*. Chicago: Gateway.

Matthews, R. 1995. *If Men Were Angels*. Lawrence: University Press of Kansas.

May, R. 1975. *The Courage to Create*. New York: Bantam.

McDonough, William, and Michael Braungart. 1998. "The Next Industrial Revolution." *Atlantic Monthly*, October, 82–92.

McKibben, W. 1998. "A Special Moment in History." *Atlantic Monthly*, May, 55–78.

McNeill, J. R. 2000. *Something New under the Sun*. New York: Norton.

Meadows, D. 1998. "The Global Citizen." *Valley News*, July 4.

Meadows, D., D. Meadows, and J. Randers. 1992. *Beyond the Limits*. Post Mills, Vt.: Chelsea Green.

Meadows, D., et al. 1972. *The Limits to Growth*. New York: Universe Books.

Merchant, C. 1980. *The Death of Nature*. New York: Harper & Row.

M'Gonigle, Michael. 1999a. "Ecological Economics and Political Ecology." *Ecological Economics* 28, no. 1: 11–26.

———. 1999b. "The Political Economy of Precaution." In *Protecting Public Health and the Environment*, ed. Carolyn Raffensperger and Joel Tickner, pp. 123–147. Washington: Island Press.

Michael, D. 1993. "The University as a Learning System." In *Universities in Crisis*, ed. W. A. W. Neilson and C. Gaffield, pp. 195–211. Montreal, Quebec: Institute for Research on Public Policy.

Midgley, M. 1985. *Evolution as a Religion*. London: Methuen.

Miller, Stephen. 1998. "A Note on the Banality of Evil." *Wilson Quarterly* 22, no. 4 (Autumn): 54–59.

Miller, W. L. 1998. *Arguing about Slavery*. New York: Vintage Books.

Minsky, M. 1994. "Will Robots Inherit the Earth?" *Scientific American* 271, no. 4 (October): 108–113.

Moravec, H. 1988. *Mind Children*. Cambridge: Harvard University Press.

Mumford, L. 1961. *The City in History*. New York: Harcourt, Brace, and World.

———. 1970. *The Myth of the Machine: The Pentagon of Power*. New York: Harcourt, Brace, Jovanovich.

Myers, N. 1997. "Consumption: Challenge to Sustainable Development." *Science* 276 (April 4): 53–57.

———. 1998. *Perverse Subsidies*. Winnipeg: International Institute for Sustainable Development.

Nabhan, G. 1982. *The Desert Smells Like Rain*. San Francisco: North Point Press.

Nabhan, Gary, and Stephen Trimble. 1994. *The Geography of Childhood*. Boston: Beacon Press.

Nadis, S., and J. MacKenzie. 1993. *Car Trouble*. Boston: Beacon Press.

National Public Radio. 2000. *Living on Earth*, May 21.

Noble, D. 1998. *The Religion of Technology*. New York: Knopf.

Norberg-Hodge, H. 1992. *Ancient Futures*. San Francisco: Sierra Club Books.

Noss, Reed. 1998a. "Sustainability and Wilderness." In *The Great New Wilderness Debate*, ed. J. Baird Callicott and Michael Nelson, pp. 408–413. Athens: University of Georgia Press.

———. 1998b. "Wilderness Recovery: Thinking Big in Restoration Ecology." In *The Great New Wilderness Debate*, ed. J. Baird Callicott and Michael Nelson, pp. 521–539. Athens: University of Georgia Press.

Oakes, J. 1998. *The Ruling Race: A History of American Slaveholders*. New York: Norton.

O'Brien, C. C. 1992. *The Great Melody*. Chicago: University of Chicago Press.

O'Brien, Mary. 2000. *Making Better Environmental Decisions*. Cambridge: MIT Press.

Ophuls, W. 1992. *Ecology and the Politics of Scarcity Revisited*. New York: W. H. Freeman.

Orr, D. 1993. "Architecture as Pedagogy." *Conservation Biology* 7, no. 2: 226–228.

———. 1994. *Earth in Mind*. Washington: Island Press.

Orwell, George. 1981. *A Collection of Essays*. New York: Harcourt, Brace, Jovanovich.

Papanek, V. 1995. *The Green Imperative*. London: Thames and Hudson.

Perrow, C. 1984. *Normal Accidents*. New York: Basic Books.

Postman, Neil. 1982. *The Disappearance of Childhood*. New York: Dell.

———. 1992. *Technopoly: The Surrender of Culture to Technology*. New York: Knopf.

———. 2000. *Building a Bridge to the Eighteenth Century*. New York: Knopf.

Press, Eyal, and Jennifer Washburn. 2000. "The Kept University." *Atlantic Monthly*, March, 39–54.

Putnam, Robert. 2000. *Bowling Alone*. New York: Simon & Schuster.

Reisberg, Leo. 2000. "10% of Students May Spend Too Much Time Online." *Chronicle of Higher Education*, June 16, A43.

Revkin, A. 2001. "Efforts to Preserve Wetlands Are Falling Short." *New York Times*, June 27, A1, A4.

Ridley, M., and B. Low. 1993. "Can Selfishness Save the Environment?" *Atlantic Monthly*, September, 76–86.

Robinson, Marilynne. 1998. "Surrendering Wilderness." *Wilson Quarterly* 22, no. 4 (Autumn): 60–64.

Rocky Mountain Institute. 1997. *Economic Renewal Guide*. Snowmass, Colo.: RMI.

———. 1998. *Green Development*. New York: John Wiley.

Rolston, Holmes. 1991. "The Wilderness Idea Reaffirmed." *Environmental Professional* 13: 370–377.

Roszak, T. 1986. *The Cult of Information*. New York: Pantheon.

Sachs, Wolfgang. 1999. *Planet Dialectics*. London: ZED Books.

Sachs, W., R. Loske, and M. Linz. 1998. *Greening the North*. London: ZED Books.

Sagoff, M. 1997. "Do We Consume Too Much?" *Atlantic Monthly*, June, 80–96.

Saul, John Ralston. 1993. *Voltaire's Bastards: The Dictatorship of Reason in the West*. New York: Vintage.

Schon, D. 1971. *Beyond the Stable State*. New York: Norton.

———. 1983. *The Reflective Practitioner*. New York: Basic Books.

Schumpeter, J. 1978. *Can Capitalism Survive?* New York: Harper Colophon.

Schwartz, E. 1971. *Overskill: The Decline of Technology in Modern Civilization*. New York: Ballantine.

Scientific American. 1989. "Managing Planet Earth." Special issue, September.

Scott, James. 1998. *Seeing Like a State*. New Haven, Conn.: Yale University Press.

Sears, P. 1964. "Ecology—A Subversive Subject." *Bioscience* 14, no. 7 (July): 11–13.

Senge, P. 1990. *The Fifth Discipline*. New York: Currency-Doubleday.

Sessions, George. 1996–1997. "Reinventing Nature? The End of Wilderness?" *Wild Earth* 6, no. 4 (Winter): 46–52.

Shepard, Paul. 1996. *Traces of an Omnivore*. Washington: Island Press.

———. 1998. *Coming Home to the Pleistocene*. Washington: Island Press.

Shepard, Paul, and Daniel McKinley, eds. 1969. *The Subversive Science*. Boston: Houghton-Mifflin.

Sherman, T. 1996. *A Place on the Glacial Till*. New York: Oxford University Press.

Shrader-Frechette, Kristin. 1993. "Locke and Limits on Land Ownership." In *Policy for Land.*, ed. Lynton K. Caldwell and Kristin Shrader-Frechette, pp. 65–83. Lanham, Md.: Rowman and Littlefield.

Shuman, Michael. 1998. *Going Local: Creating Self-Reliant Communities in a Global Age*. New York: Free Press.

Sinsheimer, Robert. 1978. "The Presumptions of Science." *Daedalus* 107, no. 2 (Spring): 23–36.

Smil, V. 1991. *General Energetics*. New York: Wiley-Interscience.
————. 1994. *Energy in World History*. Boulder, Colo.: Westview Press.
Smith, J. 1997. "The Dissenter." *Washington Post Magazine*, December 7, 18–46.
Snyder, Gary. 1995. *A Place in Space*. Washington: Counterpoint Press.
————. 1996. "Nature as Seen from Kitkitdizze Is No 'Social Construction.'" *Wild Earth* 6, no. 4: 8–9.
Somerville, Richard. 1996. *The Forgiving Air*. Berkeley: University of California Press.
Soros, G. 1997. "The Capitalist Threat." *Atlantic Monthly*, February, 45–58.
Soule, M. 1986. *Conservation Biology*. Sunderland, Mass.: Sinauer Associates.
Soule, Michael, and Gary Lease. 1995. *Reinventing Nature? Responses to Postmodern Deconstruction*. Washington: Island Press.
Spretnak, Charlene. 1997. *The Resurgence of the Real*. Reading, Mass.: Addison-Wesley.
Stegner, Wallace. [1960] 1988. "Wilderness Letter." In *Marking the Sparrow's Fall: Wallace Stegner's American West*, ed. Page Stegner. New York: Henry Holt.
Sturt, G. [1923] 1984. *The Wheelwright's Shop*. Cambridge: Cambridge University Press.
Suzuki, D. 1998. *The Sacred Balance*. Amherst, Mass.: Prometheus Books.
Thompson, D'Arcy. [1917] 1961. On Growth and Form. New York: Cambridge University Press.
Thoreau, Henry David. 1971. *The Portable Thoreau*. Ed. Carl Bode. New York: Viking.
Thornton, Joe. 2000. *Pandora's Poison*. Cambridge: MIT Press.
Todd, J., and N. Todd. 1994. *From Eco-Cities to Living Machines: Principles of Ecological Design*. Berkeley, Calif.: North Atlantic Books.
Turner, Jack. 1998. "In Wildness Is the Preservation of the World." In *The Great New Wilderness Debate*, ed. J. Baird Callicott and Michael Nelson, pp. 617–627. Athens: University of Georgia Press.
"TV Viewed as a Public Health Threat." 2001. *Environment and Health Weekly* no. 681, January 6.
Union of Concerned Scientists. 1992. "Warning to Humanity" (pamphlet). Boston.
Van der Ryn, S., and S. Cowan. 1996. *Ecological Design*. Washington: Island Press.
Veblen, T. 1973. *The Theory of the Leisure Class*. Boston: Houghton-Mifflin.
Vincent, J., and T. Panayotou. 1997. ". . . Or Distraction?" *Science* 276 (April 4): 53–55.
von Weiszacker, E., and J. Jesinghaus. 1994. *Ecological Tax Reform*. London: ZED Books.

von Weizsacker, E., A. Lovins, and H. Lovins. 1997. *Factor Four*. London: Earthscan.

Wachtel, P. 1983. *The Poverty of Affluence*. New York: Free Press.

Wackernagel, M., and W. Rees. 1996. *Our Ecological Footprint*. Philadelphia: New Society.

Wann, D. 1990. *Biologic*. Boulder, Colo.: Johnson Publishing.

———. 1996. *Deep Design*. Washington: Island Press.

Webb, W. P. 1964. *The Great Frontier*. Austin: University of Texas Press.

Weizenbaum, J. 1976. *Computer Power and Human Reason*. Cambridge: MIT Press.

White, L. 1967. "The Historic Roots of our Ecologic Crisis." *Science* 155 (March 10): 1203–1207.

Wildavsky, A. 1995. *But Is It True?* Cambridge: Harvard University Press.

Willers, Bill. 1996–1997. "The Trouble with Cronon." *Wild Earth* 6, no. 4 (Winter): 59–61.

Wilshire, B. 1998. *Wild Hunger: The Primal Roots of Modern Addiction*. Lanham, Md.: Rowman & Littlefield.

Wilson, Edward O. 1984. *Biophilia*. Cambridge: Harvard University Press.

———. 1998. *Consilience*. New York: Knopf.

Windle, Phyllis. 1994. *The Ecology of Grief*. *Orion* 13, no. 1 (Winter): 16–19.

Winner, L. 1980. "Do Artifacts Have Politics?" In *The Whale and the Reactor*, pp. 19–39. Chicago: University of Chicago Press.

World Commission on Environment and Development. 1987. *Our Common Future*. New York: Oxford University Press.

Index

Abram, D., 14, 221
Addiction, 32
Agriculture, 6, 7, 21, 22, 23, 35, 36, 37, 50, 111, 113, 138
Alexander, C., 135, 221
Amish, 5, 6, 8, 9, 46, 183, 185, 210
Anderson, E. N., 24, 221
Anderson, R., vii, 76, 77, 78, 81, 183, 221
Andropogon, Inc., 30
Applebome, P., 204, 221
Aries, P., 219, 221
Artificial intelligence, 93, 203
Ashoka Network, 157
Ausubel, J., 16, 17, 19, 20, 21, 221

Bacon, F., 156, 221
Bailey, R., 89, 221
Bali, 6, 7, 9, 35
Banuri, T., 41, 221
Bartlett, D., 115, 221
Bateson, G., 183, 222
Beck, U., 209, 222
Benyus, J., 21, 22, 139, 222
Benzing, D., vii

Berea College, 80
Berman, M., 14, 222
Berry, T., 3, 4, 107, 218, 219, 222
Berry, W., 14, 28, 42, 50, 56, 177–178, 209, 222
Bill of Rights, 212
Biomimicry, 22
Biophilia, 25, 217
Biophobia, 25
Boorstin, D., 69, 222
Boulding, K., 69, 107
Bowers, C. A., 63, 222
Brand, S., 72, 222
Braungart, M., 21, 139, 227
Browning, B., vii, 21, 130
Burke, E., 98, 99, 102, 103, 114, 212, 213, 222
Burke, K., vii
Butler, G. L., 75, 76, 77, 78, 81, 222

Caldwell, L., 100, 222
Calhoun, J. C., 144–145
California Polytechnic Institute, 160, 161, 165
Callicott, B., 190, 191, 222

Capitalism, 105–108
Capra, F., vii
Carroll, L., 35
Carson, R., 50, 76, 139, 160, 217, 222
Center for a New American Dream, 79
Chargaff, E., 35, 72, 222
Chattanooga, Tenn., 79
Churchill, W., 58, 95
Clean Air Act of 1970, 184
Climate change, 14, 16, 37, 60, 64,
   100, 111, 143, 150, 203
Coates, P., 195, 223
Cobb, C., 176, 223
Colborn, T., 200, 223
Collins, T., 139, 223
Complexity, 26
Conquest, R., 70, 223
Conservatism, principles of, 97–98
Consumption, 174, 175, 176, 177
Corporations and
   Dartmouth College case, 216
   Santa Clara County v. Southern
      Pacific Railroad, 92, 216
Cortese, A., vii
Costanza, R., 107, 214, 223
Cowan, S., 20, 21, 179, 230
Crevier, D., 93, 223
Critser, G., 201, 223
Cronon, W., 190, 191, 192, 193, 195, 223
Curitiba, Brazil, 79

Daly, H., 14, 17, 21, 107, 214, 223
Damasio, A., 95, 223
Darwin, C., 69
Deloria, V., 10, 11, 223
Devon, United Kingdom, 6, 8, 9
Diversity
   biological, 21, 23, 27, 64, 73, 134,
      203
   cultural, 27, 102
Dobb, E., 23, 223
Donahue, B., 31, 223
Dostoevsky, F., 174
Drayton, B., 157

Dubos, R., 127
Dye, N., vii

Earth Charter, 213
Easterbrook, G., 87–88, 223
Ecological design, 4, 5, 8, 9, 10, 14,
   20–32, 114, 115, 134, 179, 184
Ecological footprint, 15, 134
Ecological taxes, 21, 102
Economic growth, 11, 101, 102, 109,
   206
Edelman, M., 92, 223
Education, 31, 164–165
Ehrenfeld, D., viii, 85, 139, 223
Ehrlich, P., 120, 223
Einstein, A., 20
Ellul, J., 18, 223
Enlightenment, the, 4, 26
   ecological, 4
Environmental Protection Agency, 44
Evolution, 4, 22, 39
Ewen, S., 175, 223

Fagin, D., 177, 223
First Amendment, 215
Fishman, C., 201, 224
Foreman, D., 193, 224
Forrester, J., 25, 224
Franklin, B., 148
Franklin, C., vii, 20, 224
Free trade, 92
Full cost accounting, 131

Gandhi, M., 38
Gates, J., 115, 211, 214, 224
Gaviotas, Colombia, 79
Gelbspan, R., 144, 224
Genetic engineering, 56, 71, 110, 139
Georgescu-Roegen, N., 107
Gift economics, 10, 11
Gladwin, T., 108, 224
Goleman, D., 204, 224
Gowdy, J., 106, 107, 224
Green Mountain College, 80

Greenspan, S., 208, 224
Grumbine, E., 193, 224
Guha, R., 190, 224

Hamilton, A., 110, 112, 113
Hardin, G., 14, 67, 103, 119, 224
Harmon, A., 203, 224
Havel, V., 104, 112, 113, 180, 225
Hawken, P., 15, 20, 21, 22, 54, 73, 76,
    148, 176, 225
Hawkes, J., 8, 225
Haymount, Va., 79
Hays, C., vii
Health, 29
Healy, J., 201, 202, 204, 225
Henderson, H., 107
Henry, Mathew, 85
Heschel, A., 30, 225
Hofstadter, R., 88
Howard, A., 69
Hunter, R., 23, 225
Hunter-gatherers, 15
Huxley, A., 19, 196, 225
Hyde, L., 10, 225

Ikle, F., 102, 225
Illich, I., 182, 225
Iltis, H., 217
Industrial ecology, 21, 22, 106
Interface, Inc., 76, 79
Intergovernmental Panel on Climatic
    Change, 143, 225
Inuit, 7

Jackson, W., 21, 22, 132, 225
Jacobs, J., 181, 225
James, W., 95, 225
Jefferson, T., 47, 48, 110, 112–113, 114,
    115, 117, 148, 149, 212, 213, 225
Jesinghaus, J., 102, 230
Joy, B., 71, 203, 216, 225

Kahn, H., 16, 225
Kirk, R., 97–98, 102, 226

Kitman, J., 200, 226
Kohr, L., 51, 226
Kozol, J., 80, 226
Kuhn, T., 38, 226
Kurzweil, R., 71, 226
Kyoto Protocol, 122, 144, 151

Ladakh, 7
Land ethic, 38
Langer, S., 30, 226
Lansing, S., 7, 24, 36, 226
Lasn, K., 216, 226
Lavelle, M., 177, 223
Lawrence, D. H., 142, 196
Leach, W., 175, 226
Learning organizations, 80–81
Le Corbusier, 141
Ledewitz, B., 211, 226
Leopold, A., 17, 21, 69, 71, 75, 78, 79,
    149, 160, 178, 190, 193, 196, 226
Levins, R., 140, 226
Lewis, A. J., vii
Lewis, C. S., 16–17, 60, 226
Lewis, M., 17, 18, 226
Lewis, P., vii
Lewis Center, 132, 161, 162
Liacas, T., 216, 226
Limits to Growth, 50, 120, 227
Lincoln, A., 58
Lindblom, C., 199
Lockard, D., 215, 226
Locke, J., 100, 215, 226
Lovins, A., vii, 20, 21, 50, 73, 76, 114,
    130, 148, 226
Lovins, H., 20, 21, 50, 73, 114, 148, 226
Low, B., 94, 228
Ludwig, D. R., 121, 226
Lyle, J., vii, 21, 129, 130, 160–167, 179,
    227
Lyle Center, 161, 162, 164

Mace, S., 171–173
MacKay, A., vii
MacKenzie, J., 66

Madison, J., 47, 48, 109–110, 113, 212
Management, planetary, 17, 105
Marglin, A., 41, 221
Marsh, G. P., 13
Martin, C., 155, 227
Marx, K., 205
Maslow, A., 77, 227
Matthews, R., 109, 212, 227
McDaniel, C., vii, 106, 224
McDonough, B., vii, 21, 22, 76, 139,
   183, 227
McDonough & Partners, 130–131
McKibben, B., 15, 227
McKinley, D., 158, 229
McNeill, J. R., 146–147, 227
Meadows, D., 14, 15, 70, 120, 227
Melville, H., 63, 139
Mencken, H. L., 55
Menominee tribe, 79
Merchant, C., 14
M'Gonigle, M., 205, 227
Michael, D., 96, 227
Midgeley, T., 135, 141, 142
Midgley, M., 95, 227
Mining Law of 1872, 99
Minors Oposa, 211
Minsky, M., 93, 227
Mumford, L., 9, 14, 18, 116, 139, 227
Myers, N., 111, 115, 148, 172, 173, 227

Nabhan, G., 7, 218–219, 228
Nadis, S., 66
Nanotechnology, 19, 71
Native Americans, 10, 11
Natural capitalism, 20, 21, 73, 214
Newman, E., 55
Noble, D., 121–122
Norberg-Hodge, H., 7, 228
Normal accidents, 19, 67
Northland College, 80
Noss, R., 193, 228

Oberlin, Ohio, 43, 44, 45, 46, 118, 119,
   121

Oberlin College, 118–121, 129, 130,
   151, 161–162
O'Brien, M., 209, 228
Ocean Arks, Mass., 79
Odum, E., 21
Olmsted, F. L., 160
Ophuls, W., 183, 228
Origin of Species, 38, 50
Orwell, G., 55, 57, 61, 143, 228

Panayotou, T., 173, 230
Papago, 7
Papenek, V., 7, 228
Perrow, C., 67, 228
Political economy, 205–211
Population growth, 15, 16, 23, 51, 110,
   162
Postman, N., 19, 63, 210, 215, 228
Prairie Crossing, 79
Prescott College, 80
Press, E., 202, 228
Putnam, R., 208, 228

Quinn, D., 76

Rees, W., 15, 77, 134, 231
Regenerative design, 163–164
*Regenerative Design for Sustainable
   Development*, 164–165
Renewable energy, 21, 23, 27, 31, 115,
   148, 151
Resource efficiency, 21, 23, 100
Rethinking Progress, Inc., 79
Revelle, R., 143
Ridley, M., 94, 228
Robinson, M., 188, 189, 229
Rocky Mountain Institute, 79, 141,
   151, 172, 182, 229
Rolston, H., 190, 191, 193, 229
Roszak, T., 63, 65, 229
Ruskin, J., 171

Sachs, W., 105, 229
Safire, W., 55

Sagoff, M., 119, 120, 122, 173, 229
Saul, J. R., 20, 43, 229
Schon, D., 77, 78, 229
Schrader-Frechette, K., 215, 222, 229
Schumacher, E. F., 17
Schumpeter, J., 206, 229
Schwartz, E., 121, 229
Scott, J. C., 70, 205, 229
Sears, P., 158, 229
Senge, P., 78, 229
Sessions, G., 193, 229
Shaw, M., vii, 130
Shelley, M., 18, 63, 139
Shepard, P., 58, 207, 217, 218, 229
Sherman, T., 128, 229
Shuman, M., 115, 217, 229
Silent Spring, 50, 76, 139
Sinsheimer, R., 215, 216, 229
Slavery, 144–148
Smil, V., 14, 15, 230
Smith, A., 179, 205
Social traps, 38
Sontag, S., 60
Soros, G., 107, 230
Soule, M., 58, 193, 230
Spirn, A. W., 192, 193
Steele, J., 115, 221
Stegner, W., 187, 230
Stranahan, M., vii
Strong, S., vii
Sturt, G., 10, 230
Subsidies, 66, 111, 115, 148, 184
Sustainability, 11, 50, 92, 94, 95, 173, 195
Swift, J., 195
Systems dynamics, 178, 205

Technocracy, 19
Technopoly, 19, 31
Thompson, D., 21, 230
Thoreau, H. D., 178, 190, 230

Thornton, J., 137, 199, 230
Todd, J., vii, 28, 130, 230
Todd, N., 28
Trimble, S., 218–219
Tufts University, 151
Tuluca, A., vii
Turner, J., 188, 230

Union of Concerned Scientists, 140, 230

Valery, P., 68
Van der Ryn, S., 20, 21, 179, 230
Veblen, T., 175, 230
Village Homes, 79
Vincent, J., 173, 230
Virginia Bill of Rights, 212
Von Weiszacker, E., 102, 230, 231

Wachtel, P., 176, 231
Wackernagel, M., 15, 134, 177, 231
Wann, D., 20, 179, 231
Warren Wilson College, 80
Washburn, J., 202, 228
Waste, 15, 27, 134, 176, 178, 182, 209
Webb, W. P., 109, 231
Weizenbaum, J., 62, 66, 231
Wetlands, loss of, 44–45
White, L., 13, 231
Wildavsky, A., 89, 231
Willers, B., 193, 231
Wilshire, B., 32, 231
Wilson, E. O., 18, 19, 25, 217, 231
Windle, P., 207, 231
Winner, L., 121, 231
Wise use movement, 85
World Resources Institute, 79
WorldWatch Institute, 79
Wright, F. L., 136

Yup'ik, 155